Acid Stress
and
Aquatic Microbial
Interactions

Editor

Salem S. Rao, Ph.D.
Research Scientist
National Water Research Institute
Canada Centre for Inland Waters
Burlington, Ontario, Canada

CRC Press, Inc.
Boca Raton, Florida

Library of Congress Cataloging-in-Publication Data

Acid stress and aquatic microbial interactions / editor, Salem S. Rao.
     p.   cm.
   Bibliography: p.
   Includes index.
   ISBN 0-8493-5168-5
   1. Freshwater microbiology.   2. Acid rain--Environmental aspects.
3. Microbial ecology.   I. Rao, S. S. (Salem S.), 1934-
QR105.5.A19 1989
576'.15--dc19   88-22257
                                                CIP

Direct all inquiries to CRC Press, Inc., 2000 Corporate Blvd., N.W., Boca Raton, Florida, 33431.

© 1989 by CRC Press, Inc.

International Standard Book Number 0-8493-5168-5

Library of Congress Number 88-22257
Printed in the United States

PREFACE

During the last 2 decades, emphasis has been primarily on studies aimed towards acquiring an understanding of the interrelationships between microbial communities and environments experiencing acid stress. This is a consequence of the role played by microbes in the biogeochemical processes and energy flow in these ecosystems.

Ecological interactions have been defined as the study of the relation of organisms to their environment. Many, if not all, of the important transformations occurring in an ecosystem are due to the presence and activities of microbes. Nevertheless, considerable research is still required to clearly understand and evaluate the concepts and principles of ecological interactions at the microbial level. Environmental processes such as those responsible for the adaptation of microbial communities to extreme environments in order to bring about key microbiological reactions are not fully understood.

This volume presents information and techniques which are at the leading edge of microbial acid rain research and addresses a number of topical and important issues of global concern. These are microbial response to low pH, biogeochemical processes, cycling of organic matter, microbial interactions with higher forms of biota, and the factors affecting the above processes and interactions. This book should be of special interest to researchers and those involved in management of acid-stressed waters.

THE EDITOR

Salem S. Rao, Ph.D., a specialist in microbiology of lakes and rivers, has conducted an extensive research program for over 10 years to study the effects of acid stress on lake microbiological processes and mechanisms of lake recovery. As an expert, he has presented many international seminars on this topic. He has authored and co-authored over 80 papers and reports on such topics as bacterial populations and processes in the Great Lakes and acid-stressed lakes, role of bacteria in the oxygenation and deoxygenation processes, psychrophilic bacterial activity, bacterial-phosphate kinetics, sediment microbiology, ecotoxicological assessment of sediments using microbial community structure, and sediment microbial processes.

Dr. Rao has been chairman of two task groups in ASTM Committee D:19:24 on water microbiology. He is also the co-chairman of a symposium committee which organizes the International Symposium on Aquatic Microbiology Series. He holds associate positions in Brock University and the University of Toronto.

CONTRIBUTORS

Pamela Bell, Ph.D.
Research Associate
Department of Microbiology
University of Virginia
Charlottesville, Virginia

Ronald Benner, Ph.D.
Environmental Scientist
Environmental Research Laboratory
Environmental Protection Agency
Athens, Georgia

B. Kent Burnison, Ph.D.
Research Scientist
Labes Research Branch
National Water Research Institute
Burlington, Ontario, Canada

Mike Dickman, Ph.D.
Professor
Department of Biological Sciences
Brock University
St. Catharines, Ontario, Canada

Sushil S. Dixit, Ph.D.
Research Scientist
Department of Biology
Queen's University
Kingston, Ontario, Canada

Jerry W. Elwood, Ph.D.
Research Staff Member
Environmental Sciences Division
Aquatic Ecology Section
Oak Ridge National Laboratory
Oak Ridge, Tennessee

Alan T. Herlihy, Ph.D.
Research Assistant Professor
Utah Water Research Laboratory
Utah State University
Environmental Protection Agency
Corvallis, Oregon

Robert E. Hodson, Ph.D.
Professor
Department of Microbiology
and
Institute of Ecology
University of Georgia
Athens, Georgia

David L. Lewis, Ph.D.
Research Microbiologist
Environmental Protection Agency
Athens, Georgia

Aaron L. Mills, Ph.D.
Associate Professor
Department of Environmental Sciences
University of Virginia
Charlottesville, Virginia

Myron J. Mitchell, Ph.D.
Professor
Department of Environmental Forestry
  Biology
SUNY/CESF
Syracuse, New York

MaryAnn Moran, Ph.D.
Postdoctoral Associate
Department of Microbiology
University of Georgia
Athens, Georgia

Patrick J. Mulholland, Ph.D.
Research Staff Member
Environmental Sciences Division
Oak Ridge National Laboratory
Oak Ridge, Tennessee

Robert E. Murray, Ph.D.
Department of Biological Sciences
University of Kentucky
Lexington, Kentucky

Jeffrey S. Owen
Research Assistant
Department of Environmental Forestry
  Biology
SUNY/CESF
Syracuse, New York

Anthony V. Palumbo, Ph.D.
Research Associate
Environmental Sciences Division
Oak Ridge National Laboratory
Oak Ridge, Tennessee

Salem S. Rao, Ph.D.
Research Scientist
Biological Department
Brock University
St. Catharines, Ontario, Canada

John P. Smol, Ph.D.
Assistant Professor
Department of Biology
Queen's University
Kingston, Ontario, Canada

Jeffrey D. Teska, Ph.D.
Department of Medical Microbiology
University of Georgia
Athens, Georgia

Sarah C. Tremaine, M.D.
Graduate Student
Department of Environmental Sciences
University of Virginia
Charlottesville, Virginia

TABLE OF CONTENTS

Chapter 1

MICROBES, SEDIMENTS, AND ACIDIFIED WATER: THE IMPORTANCE OF BIOLOGICAL BUFFERING

Aaron L. Mills, Pamela E. Bell, and Alan T. Herlihy

TABLE OF CONTENTS

I. INTRODUCTION

Acid pollution affects a large number of water bodies in North America and northern Europe. While the most commonly referenced pollutant is acid precipitation, intense acidification of lakes and streams often arises from acid mine drainage (AMD).[1,2] Although it is clear that acid pollution causes a great deal of damage to receiving waters (i.e., fishkills and oligotrophication), it is also clear that there are a number of reactions in the water column and sediments than can help to combat the acidification. A number of chemical reactions are of importance, including mineral weathering and ion exchange with colloidal materials such as clays. These reactions allow a catchment to neutralize some of the acidity in precipitation or mine drainage before the incoming water arrives in the receiving stream or lake. The buffering due to reserves of base cations and weatherable materials is in limited supply in many watersheds and completely absent in others, so that the response of watersheds to acidification is frequently termed "delayed", that is, the acid is being stored in the soils of the catchment temporarily, but will eventually be released into the water.[3]

Lake 223 in the experimental lakes area was acidified by adding a known amount of sulfuric acid over a period of several years to the epilimnion. Schindler et al.[4] found that the pH observed in Lake 223 was much greater than the pH predicted based on the quantity of acid added to the system. They found that 62.5 to 69.1% of the acid introduced into the lake was neutralized by chemical species other than dissolved bicarbonate present in the water column. This was termed auxiliary buffering and was due mostly to biologic sulfate reduction (SR) occurring in the sediments of Lake 223.

A number of biological reactions occur that can neutralize acidity.[5-10] These reactions include photosynthesis, denitrification, iron reduction, ammonification, and SO_4^{2-} reduction. In each case, the neutralization of the protons or protogenic species may be temporary or permanent, depending on whether the products of the reactions are available for further acidogenic metabolism. For example, the generation of ammonia during organic matter decomposition (ammonification) can neutralize acidity by the reaction

$$NH_3 + H^+ \rightarrow NH_4^+ \tag{1}$$

However, the ammonia thus generated is readily oxidized by chemoautotrophic bacteria to form nitrate with the release of the proton

$$NH_4^+ + 3/2O_2 \rightarrow H^+ + NO_3^- \tag{2}$$

resulting in no net neutralization of acidity.

Because most acidification of current concern is caused by sulfate aerosols resulting in the deposition of acid sulfate (sulfuric acid), much of the research on biological buffering has examined the role of SR in the neutralization of acidity. This reaction constitutes a true homeostatic mechanism for acid neutralization, in that the process is enhanced by deposition of the pollutant itself and tends to move the pH of the water toward a "set point" that would represent neutral conditions.[11] SR is not the only microbial reaction leading to alkalinity generation; however, each of the others listed above plays some role in neutralizing acidity, and in some cases one or more of the reactions named may exceed SO_4^{2-} reduction in importance to the acid-neutralizing capacity (ANC) of a particular water body.

Before detailing how bacteria function in neutralization of acidity, it is important to distinguish between alkalinity and pH. pH is an intensity factor that describes the H^+ concentration (activity). Alkalinity, or ANC, describes the excess of positive charges over the anions of strong bases. Alkalinity is operationally defined as the amount of a strong acid necessary to titrate a solution to pH 4.5. This represents the inflection point of a carbonate titration curve. Carbonate alkalinity is defined as

$$\text{Alk} = [HCO_3^-] + 2[CO_3^{2-}] + [OH^-] - [H^+] \tag{3}$$

However, most water contains pH-reactive species other than carbonate, and microbial metabolism may have a significant impact on many of these ions. We can expand the alkalinity equation (Equation 3) to include contributions from other dissolved species as well as organic molecules and ammonium ions:[12]

$$\text{Alk} = [HCO_3^-] + 2[CO_3^{2-}] + [RCOO^-] + [OH^-] - [H^+]$$
$$= [Na^+] + [K^+] + 2[Ca^{2+}] + [NH_4^+] + 2[Mg^{2+}] + 2[Fe^{2+}]$$
$$+ 3[Fe^{3+}] - [Cl^-] - 2[SO_4^{2-}] - [NO_3^-] - [org^-] \tag{4}$$

In order to maintain charge and proton balance it is preferable to choose uncharged species as a zero-proton reference state, e.g., RCOOH (undissociated organic acids), NH_3, H_3BO_3, CO_2, etc. Thus, NH_4^+ is a cation of a strong base and is included as such in the charge balance equation.

A change in alkalinity must be accompanied by a change in anions and cations other than H^+ to preserve electroneutrality. Berner et al.[7] suggested that the major in-lake processes affecting alkalinity are

1. Calcium carbonate dissolution/precipitation

$$CaCO_3 + Ca^{2+} + H_2O = Ca^{2+} + HCO_3^- \tag{5}$$

$$CaCO_3 = Ca^{2+} + CO_3^{2-} \tag{6}$$

2. Authigenic silicate formation

$$3Al_2Si_2O_5(OH)_4 + 2K^+ + 2HCO_3^- = 2KAl_3Si_3O_{10}(OH)_2 + 2CO_2 + 5H_2O \tag{7}$$

Silicate formation may also occur with Na^+ or Mg^{2+} as the cation.

1. SR

$$2CH_2O + SO_4^{2-} = S^{2-} + 2CO_2 + 2H_2O \tag{8}$$

$$S^{2-} + 2CO_2 + 2H_2O = H_2S + 2HCO_3^{2-} \tag{9}$$

2. Ammonia formation

$$CH_2NH_2COOH + 2(H) = NH_3 + CH_4 + CO_2 \tag{10}$$

$$CO_2 + NH_3 + H_2O = NH_4^+ + HCO_3^- \tag{11}$$

$$NH_3 + HCO_3^- = NH_4^+ + CO_3^{2-} \tag{12}$$

II. MICROBIAL COMMUNITIES IN ACIDIFIED SYSTEMS

In order for microorganisms to have any impact on acidic pollution, those organisms must be able to survive and grow in that environment. Understanding of the ability of microbes to assist in neutralizing acidity comes from an understanding of how microbial communities function in acidified waters and sediments.

A. Water

Although there are numerous articles in the literature concerning the effect of pH on microorganisms in laboratory situations, the pH effect on communities in the environment is, within limits, less severe. Microbial response to extremely acid conditions has been studied at Lake Anna, VA. The contrary creek arm of this impoundment receives AMD from a series of abandoned pyrite mines. The acid mine stream itself has a pH of about 2.5, with concentrations of sulfate of 2 to 20 mM, and dissolved iron concentrations of 10 to 50 mg/l. The pH of the water in the lake receiving the mine drainage ranges from 3.2 at the mouth of Contrary Creek to above 6.0 at a point about 2 km down the arm. Data and examples from the Lake Anna studies will be used extensively throughout this discussion as examples of the behavior of microorganisms in very acid situations.

In the waters of Lake Anna, leaf litter decomposition is known to be reduced under acidic conditions,[13] and the activity of the bacterioplankton is similarly reduced,[14] although neither change is proportional to the degree of acidification. Furthermore, the community structure of planktonic communities in acidified waters varies from communities in circumneutral waters. In their study of bacterioplankton in acidic waters, Wassel and Mills[14] suggested that the suspended community suffered from the acid stress because the organisms were transients. They suggested that communities of microbes attached to stationary surfaces would represent the true indigenous community and that such a community might function at a higher level in the acidic water.

Mills and Mallory[15] addressed the question raised earlier by Wassel and Mills.[14] Microbial communities developed on glass slides suspended in acid-polluted Contrary Creek (pH = 2.9) and nonpolluted Freshwater Creek (pH = 6.5) but otherwise similar waters showed evidence of stress when suspended at the opposite station. Glucose incorporation was inhibited in both translocated communities, but the inhibition was not as severe and recovery of activity was faster for the acid-developed community as compared to the circumneutral community. Numerical taxonomy of the communities showed that they contained a substantially different set of members with little overlap. The range of pH values at which the members of the acid-developed community could function suggested that the members of that community were generalists as opposed to narrowly constrained members of the community from the circumneutral station. Based on the proportion of test characters that received positive responses, the organisms from the acidic site were more general in their abilities (47.6% positive) as compared with the neutral counterparts (18.7% positive). The results support the concept that communities developed in extreme environments tend to be generalists whereas those from mesic environments, due to the higher levels of competition present, tend to be specialists. Clearly, heterotrophic organisms are capable of developing fully functional communities in acidified waters.

B. Sediments

While aerobic heterotrophic communities suffer in acidified systems, anaerobic communities may actually thrive. There is less competition for carbon from the aerobes and more carbon may be delivered to the sediments in acid-impacted water. We compared anaerobic heterotrophs by the most-probable-number technique and found that in Contrary Creek there are more anaerobic heterotrophs than in Freshwater Creek (Table 1).

Most of the acidified freshwater of the world contains higher than "average" sulfate concentrations. Current theories of competition between sulfate reducers and methanogens suggest that when sulfate is not limiting to sulfate reducers, they will outcompete methanogens for the common energy source ($H_2(g)$).[16-19] If sulfate reducers can outcompete methanogens for their energy source, then the growth rate of sulfate reducers should increase and they should outnumber methanogens.

The methanogen (CA) and sulfate reducer (SM) were counted with depth along a gradient

Table 1
ABUNDANCE OF ANAEROBIC BACTERIA IN THE SEDIMENTS OF THE CONTRARY CREEK ARM (ACIDIC) AND FRESHWATER CREEK ARM (UNPOLLUTED) OF LAKE ANNA, VA

	Contrary Creek	Freshwater Creek
Anaerobic Heterotroph (MPN)	$1.3 \cdot 10^6$ $(3.94 \cdot 10^5 - 4.29 \cdot 10^6)^a$	$1.1 \cdot 10^5$ $(3.33 \cdot 10^4 - 3.63 \cdot 10^5)^a$
Methanogen (MPN)	$1.3 \cdot 10^5$ $(3.94 \cdot 10^4 - 4.29 \cdot 10^5)^a$	$4.9 \cdot 10^4$ $(1.48 \cdot 10^4 - 1.62 \cdot 10^5)^a$
Methanogen (FA)	$3.62 \cdot 10^6$ $(1.58 \cdot 10^6)^b$	$4.01 \cdot 10^6$ $(1.65 \cdot 10^6)^b$

[a]   Represents 95% confidence interval.
[b]   Represents SEM.

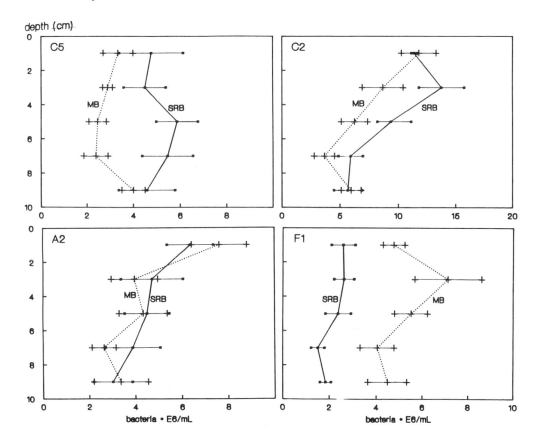

FIGURE 1.   Fluorescent antibody counts of a methanogen and SRB isolated from the sediments of the Contrary Creek arm of Lake Anna. Bars represent ± one SEM. Stations sampled are along a gradient of water column sulfate concentrations in the order C2 > C5 > A2 > F1, where F1 is a freshwater station that does not receive any acid sulfate and C2 is the mouth of a creek that received acid mine drainage. Depth in centimeters.

of column water sulfate concentrations in the Contrary Creek arm of Lake Anna using an indirect fluorescent antibody assay.[20] Triplicate cores were taken at each station and duplicate samples were counted for each core (Figure 1). The relative sulfate gradient of the stations is C5 > C2 > A2 > F1. Where the water column concentration of sulfate is the highest (station C5), sulfate reducers significantly outnumber the methanogens and, where water

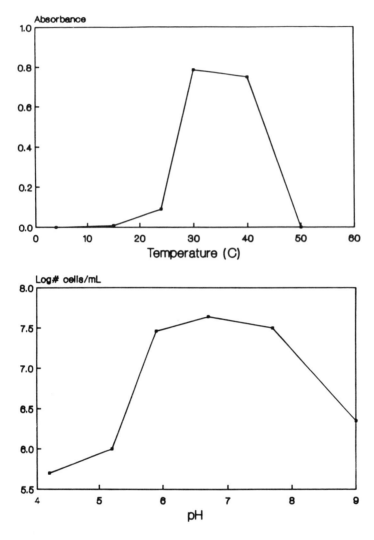

FIGURE 2.   (above) pH response of a sulfate reducer, *Desulfovibrio* strain SM, isolated from the sediments of the Contrary Creek arm of Lake Anna. (below) Temperature response for the same organism as above. Both values are taken after 24 h of growth.

column sulfate concentrations are lowest (station F1), methanogens significantly outnumber sulfate reducers. Station C2 still has more sulfate reducers and station A2 has equal numbers of both groups. While these counts certainly do not represent the whole guild of sulfate reducers or methanogens, MPN comparisons with the fluorescent antibody indicate that the antibody count gives one-two orders of magnitude higher numbers than does the MPN technique (Table 1) and we feel that the numbers obtained by the fluorescent antibody method are a good estimate of the guild.

Many have suggested that organisms living in the sediments underneath acidified water must be acid tolerant, and Konopka et al.[21] have isolated sulfate-reducing bacteria that show pH optima at acidic pH. The strain of *Desulfovibrio vulgaris* isolated from the sediments of Contrary Creek has a pH optimum of 6.8 and a temperature optimum between 30 and 40°C (Figure 2). It is important to note that the pH of sediments underlying the Contrary Creek arem are at pH 6.5 within 1 cm of the surface, even when the pH of the overlying water is below 4.0.[22]

Table 2
OXYGEN CONSUMPTION BY
BACTERIA FROM SEDIMENT
COMMUNITIES MIXED WITH
WATER FROM THE SAME OR
OPPOSITE SITE

Source of sediment organisms	Source of water for incubation	Replicate	
		A	B
CC	CC	31.9	36.9
CC	FC	60.1	75.3
FC	CC	184.8	13.9
FC	FC	365.6	753.6

Note: Values are microliters O_2 taken up in 24 h.

To assess the health of the aerobic heterotrophic community and the impact of allochthonous water on an aerobic community, oxygen uptake was measured with Contrary Creek and Freshwater Creek surface sediment bacteria in their *in situ* water as well as in the water from the opposite creek using a Warburg respirometer. Of a 10^9 cells per milliliter overnight enrichment, 0.1 ml was inoculated into each of several Warburg flasks along with 1.8 ml of filter sterilized lake water containing 1 mM filter sterilized glucose. KOH was placed in the center well to absorb any CO_2 produced during the experiment. The organisms from Freshwater Creek incubated in their *in situ* water respired the most (Table 2) while the organisms from Contrary Creek incubated in their *in situ* water respired the least. The cross-inoculation experiments revealed that Freshwater Creek water increased the O_2 uptake by Contrary Creek cells and Contrary Creek water depressed respiration by the Freshwater Creek cells.

III. BIOLOGICAL REACTIONS THAT INCREASE PH AND/OR ALKALINITY

While we acknowledge that glucose is a source of carbon in anaerobic systems, as conditions become more reducing, the pool of low molecular weight organic acids and alcohols increases and the direct role of glucose as a source of carbon and energy decreases. The use of organic anions instead of $(CH_2O)_n$ in equilibrium equations makes a large difference in the conclusions that may be drawn concerning the role of biochemical reactions on alkalinity generation and proton consumption. We will more closely approximate *in situ* equilibrium conditions by using organic anions as carbon sources under highly reducing conditions instead of the generic CH_2O.

A. Photosynthesis
Photosynthesis occurs according to

$$nCO_2 + nH_2O = (CH_2O)_n + nO_2 \tag{13}$$

There is no net change in alkalinity resulting from this reaction through the local pH will increase due to the removal of CO_2. When photosynthesis occurs with the assimilation of nitrate according to

$$106CO_2 + 16NO_3^- + HPO_4^{2-} + 122H_2O + 18H^+ = (C_{106}H_{263}O_{110}N_{16}P) + 138O_2 \tag{14}$$

alkalinity will increase due to the removal of protons (Equation 4). If the charge balance form of the alkalinity equation is used, then alkalinity is seen to increase due to removal of nitrate and phosphate anions.

B. Nitrate Reduction

Denitrification is the reduction of nitrate with dinitrogen gas as the major product and results in the consumption of protons:[23]

$$\text{glucose} + 4.8NO_3^- + 4.8H^+ = 6CO_2 + 2.4N_2 + 8.4H_2O \tag{15}$$

Respiration of nitrate with nitrite as a product will result in a decrease in pH due to the production of CO_2 but no change in alkalinity:

$$\text{glucose} + 12NO_3^- = 6CO_2 + 6H_2O + 12NO_2^- \tag{16}$$

When nitrate is respired with ammonia as a product alkalinity is produced:

$$\text{glucose} + 3NO_3^- + 3H^+ = 6CO_2 + 3NH_4^+ + 3OH^- \tag{17}$$

C. Manganese and Iron Reduction

Minerals of manganese and iron such as amorphous manganese oxyhydroxides, manganite (MnOOH), pyrolusite (MnO_2), amorphous iron oxyhydroxides, hematite (Fe_2O_3), and goethite (FeOOH) are common sources of iron and manganese available for biologic reduction. Inorganic dissolution of minerals tends to produce alkalinity,[24] and we will explore the impact of biochemistry on alkalinity generation or proton removal.

Reduction of both iron and manganese may occur as a direct or indirect result of microbial metabolism.[25-27] Equations 18 and 19 assume direct microbial reduction with a disproportionation of glucose to lactate + $3CO_2$. The reduction of manganese may occur according to:

$$MnO_2 + \text{glucose} =$$
$$Mn^{+2} + CH_3CHOHCOO^- + 3CO_2 + 3/2H_2 + OH^- + 2H_2O \tag{18}$$

For iron reduction:

$$FeOOH + \text{glucose} + H_2O =$$
$$Fe^{+2} + CH_3CHOHCOO^- + 3CO_2 + 3.5H_2 + 5H_2 + OH^- \tag{19}$$

When iron and manganese reduction reactions are written using organic acids as carbon sources, more alkalinity and less CO_2 is generated than from the oxidation of glucose:

$$MnO_2 + CH_3CHOHCOO^- + H_2O =$$
$$Mn^{2+} + CH_3COO^- + CO_2 + 2OH^- + (H) \tag{20}$$

$$FeOOH + CH_3CHOHCOO^- + H_2O =$$
$$Fe^{+2} + CH_3COO^- + CO_2 + H_2 + 2OH^- \tag{21}$$

Thus, the complete oxidation of a mole of glucose to CO_2 with iron or manganese as the terminal electron acceptor can neutralize 3 mol of protons. Both of these reactions result in the generation of temporary alkalinity if the metal cation is not retained within the system. The oxidation of these metals results in the generation of acidity (protons) due to the splitting of water to form oxidized mineral phases.

D. Amino Acid Fermentation

Amino acid fermentation may occur by using either one or two amino acids as carbon and energy sources as well as terminal electron acceptor. The products of amino acid fermentation are NH_3, CO_2, H_2S, fatty acids, and other volatile substances. For example, a Stickland fermentation uses pairs of amino acids, one as the electron donor, and one as the electron acceptor.

$$\text{alanine} + 2\text{ glycine} + 2H_2O = 3NH_4^+ + 3CH_3COO^- + CO_2 \tag{22}$$

Production of organic acids increases alkalinity as does the production of ammonium ion (Equation 4). Again, CO_2 production has no effect on alkalinity, since for each CO_2 that dissolves, one proton is produced for each bicarbonate anion generated. Fermentation of a single amino acid results in the production of alkalinity from pyruvate or ammonium according to:

$$\text{serine} = \text{pyruvate}^{(-)} + NH_4^+ \tag{23}$$

E. Methanogenesis

Methanogenesis occurs in the sediments underlying most (probably all) bodies of water. This reaction results in the complete reduction of CO_2 to CH_4 using H_2 as an energy source:

$$CO_2 + 4H_2 = CH_4 + 2H_2O \tag{24}$$

There is no change in alkalinity from this reaction though the pH may increase due to the reduction in PCO_2.

F. Sulfate Reduction

SR has been written many ways and is generally stated to increase alkalinity according to the reaction:[7,28]

$$2CH_2O + SO_4^{2-} = S^{2-} + 2CO_2 + 2H_2O \tag{25}$$

$$S^{2-} + 2CO_2 + 2H_2O = H_2S + 2HCO_3^- \tag{26}$$

which results in the combined reaction

$$2CH_2O + SO_4^{2-} = H_2S + 2HCO_3^- \tag{27}$$

Equation 27 describes the net generation of alkalinity by sulfate reduction. Sulfate-reducing bacteria (SRB) use many different fatty acids and alcohols as sources of carbon when reducing sulfate, though a few strains have been described that oxidize glucose.[29] These bacteria probably release $H_2S(g)$ as a product of respiration and not S^{2-} or HS^-. When these two facts are taken into account the sulfate reduction equation becomes:

$$SO_4^{2-} + CH_3CHOHCOO^- + H_2 = H_2S(g) + CO_2 + CH_3COO^- + 2OH^- \tag{28}$$

This reaction results in the net generation of alkalinity if the $H_2S(g)$ escapes from the system and does not become reoxidized. If H_2S remains within the system and dissociates into $H^+ + HS^-$, then only 1 mol of hydroxide will be generated for each mole of sulfate reduced since the proton generated from sulfide dissociation will react with hydroxyl to form water. The pK_1 of H_2S is 7.0; at pH = 7 H_2S and HS^- will be present in solution in equal concentrations. When the pH is 6, the concentration of HS^- is one tenth that at pH 7. This means that when the $H_2S(g)$ dissoves in water and dissociates according to its thermodynamic constraints, the lower the pH of the water (below 7) the more alkalinity will be generated by sulfate reduction.

1. Measuring Rates of Bacterial Sulfate Reduction

Rats of SR in the environment have been measured by several means including the use of radiolabeled sulfate tracers, with enclosure experiments, or by utilizing pore water sulfate concentration depth profiles and diagenetic models. Measuring SR rates by determining the amount of reduced sulfur in the sediment is not a reliable estimate since a large fraction of the reduced sulfur can be oxidized at the anoxic/oxic interface.[30] Bacterial counts of sulfate reducers are often given as evidence for sulfate reduction, however, at best, such counts provide only an indication of the potential for SR to occur. Only direct measurement (or calculation) of the rate by one of the following methods can provide a true rate of reduction and allow computation of alkalinity from the reaction.

a. Radioactive Tracer Technique

The most widely used method to measure SR rates involves the use of ^{35}S, a low energy, β-emitting sulfur isotope. ^{35}S has been used to measure SR rates since the early 1960s by a large number of researchers.[31-33] Jørgensen[34] describes in detail the methodology involved in using ^{35}S to measure SR in sediments. In this technique small amounts (<50 μl) of carrier-free ^{35}S-SO_4^{2-} are injected into the sediment or water sample to be measured and the samples are incubated at the *in situ* temperature for an appropriate amount of time. The incubation is terminated, usually by quick freezing, and the amount of labeled reduced sulfur formed over the incubation period is determined. The sulfate turnover rate constant (k_{SR}) is then calculated by Equation 29.

$$k_{SR} = {}^{35}S\text{-reduced}/({}^{35}SO_4^{2-} \text{ added}/t) \tag{29}$$

where t is the incubation time. The turnover rate constant is thus a ratio indicating the fraction of the sulfate molecules reduced per unit time. To calculate the SR rate it is necessary to know the sulfate concentration in the sample (sediment or lake water).

$$\text{SR rate} = k_{SR} * [SO_4^{2-}] * \alpha \tag{30}$$

where α is the isotope fractionation factor which represents the preference of the SRB for the lighter sulfur isotopes (^{32}S and ^{34}S) as opposed to ^{35}S. It has been shown that ^{32}S-SO_4^{2-} was metabolized 2 to 4% faster than ^{34}S-SO_4^{2-},[28] thus the value of the isotope fractionation factor for ^{35}S ranges from 1.03 to 1.06.[34]

The basis of the radiotracer method is that the proportion of labeled sulfate molecules reduced reflects the overall proportion of unlabeled sulfate molecule reduced. Thus, even though the distribution of the label will not be uniform throughout the sample, the measurement of SR is valid since at each point at any given moment, the same fraction of labeled sulfate is reduced per unit time as unlabeled sulfate (corrected for isotope fractionation).

Almost all of the ^{35}S work has been done in marine/salt marsh systems with high concentrations of sulfate. In applying this technique to freshwater systems, extra care must be taken in determining the incubation time and the amount of label used in the determination.

The low sulfate concentrations in freshwater systems mean that there is a danger of adding label in amounts that would cause an increase in the total sulfate concentration in the sample thus causing artificially high rates of SR. Therefore, only small amounts of high-specific activity label should be used in low-sulfate samples and calculations done to insure that tracer amounts (<10% of ambient concentrations) of ^{35}S are added. It is also critical that time series experiments be performed in each system to be measured to determine the appropriate incubation time. It is necessary that the rate of sulfate turnover be linear between time zero and the selected incubation time. In freshwater systems sulfate turnovers are much more rapid than they are in marine systems. Between 20 and 90% of the added sulfate label has been shown to be reduced within 1 h in low-sulfate freshwater lake sediments.[35,36]

In most of the older literature, the amount of reduced sulfur formed in the sample was determined by acidifying the samples and trapping the liberated sulfide in NaOH or zinc acetate. This methodology only detected acid-volatile sulfur (H_2S and amorphous FeS). Recent studies have shown that a large fraction of the added sulfate label can be reduced into non-acid-volatile forms such as pyrite, elemental sulfur, or organic sulfur (ester sulfates or carbon bonded sulfur).[37-39] Thus, it is necessary that the sample be analyzed for the presence of the ^{35}S label in the full spectrum of sulfur species (see References 40 and 41 for descriptions of sulfur analyses).

Isotope exchange is a potential problem with the isotope tracer technique.[37] If the ^{35}S atoms were to exchange between sulfate and sulfide (without a chemical transfer between sulfate and sulfide), this technique would be invalid. Although there has been no conclusive work reported in the literature showing isotope exchange between sulfate and sulfide, some evidence, however, suggests that isotope exchange does occur between sediment sulfur species.[42] This problem needs to be addressed in future work with this technique.

b. Enclosure Experiments

Another way to measure SR is to set up sediment/lake water enclosures and to monitor the depletion of sulfate with time. This process has been used on a scale ranging from small sediment microcosms[43] to large-scale lake enclosure mesocosms[44] to whole lake studies treating the lake hypolimnion as an enclosure during summer stratification.[4,45] In reality, all of these enclosures are measuring sulfate removal rates and not SR rates. Sulfate removal measures the amount of sulfate removed from the lake water by SR minus the amount of sulfate generated during sulfide oxidation. If sulfide oxidation is not occurring or is negligible, then the sulfate removal rate is equal to the SR rate. Other processes, e.g., adsorption of SO_4^{2-} to iron-rich particles, can also contribute to the measured sulfate retention, depending on the methods used for measuring SO_4^{2-} removal. Conservative alkalinity is equal to the SO_4^{2-} removal. The enclosure method is appropriate for determining amounts of sulfate removed over long-term (days to months) time intervals. In order to extrapolate actual *in situ* instantaneous rates of SR, care must be taken to insure that the rate of sulfate depletion in the enclosure is linear with time. The smaller scale enclosures are also susceptible to the so-called "bottle effects".

c. Diagenetic Modeling

Bacterial SR rates have been estimated based on sediment depth profiles of sulfate and a theoretical model of sulfate diagenesis.[46]

$$\frac{\partial c}{\partial t} = D_s \frac{\partial^2 c}{\partial x^2} - w \frac{\partial c}{\partial x} - a * e^{(-bx)} \qquad (31)$$

where c = concentration, t = time, D_s = diffusion coefficient, x = depth in the sediment (positive in the downwards direction), w = sedimentation rate, b = rate constant for SR, and a = rate constant for SR.

The term containing D_s represents sulfate diffusion into the sediments and the term containing w represents convective transport of sulfate to the sediment via deposition and compaction. The $a*e^{(-bx)}$ term was derived by Berner[47] to model SR and represents the first-order consumption of sedimenting organic matter by the SRB. By assuming steady-state (dc/dt = 0) and two boundary conditions (at x = 0, c = c_o and at x = infinity, c = c_{inf}) Equation 31 can be solved for the concentration of sulfate at any depth.

$$C = (C_o - C_{inf}) * e^{(-bx)} + C_{inf} \qquad (32)$$

$$(C_o - C_{inf}) = a/(D_s * b^2 + w * b) \qquad (33)$$

The observed sulfate depth profile is fitted to an exponential function of the form:

$$C = A * e^{(-Bx)} + C \qquad (34)$$

The fitted value of A would then be equal to $(C_o - C_{inf})$, C is equal to C_{inf}, and B is equal to the rate constant b. Then by measuring D_s and w, the rate constant a can be calculated and from knowledge of a and b the depth distribution of SR can be calculated from Equation 35.[48]

$$dc/dt = a * e^{(-bx)} \qquad (35)$$

There are many assumptions involved in this technique. One is that the system is at steady state with respect to both deposition and consumption of organic matter. The second assumption is that both b and D_s are constant with depth. It is also assumed that there is negligible carbon utilization by other bacteria and that there is no SR below the c_{inf} depth used to fit the model. These assumptions make the use of this diffusion modeling method of measuring SR of little value in lakes. The assumption of steady state in lakes is almost always going to be violated. Also, the factor limiting SR in freshwater lakes (see Section 2) is usually sulfate concentration and not organic matter concentration.

2. Factors Controlling Sulfate Reduction Rates

SR requires sulfate, organic substrates, an appropriate temperature, and anoxic conditions. Thus, major factors controlling the capacity of a lake to neutralize incoming acidity are the amounts of organic matter input, sulfate concentration, and lake hydrology. Physical processes such as diffusion and lake hydrology determine the rate of supply of reactants to the anoxic regions of the lake where SR occurs. Once there, it is chemical factors (concentrations of sulfate and organic substrates) that control the overall rate of SR.

a. Chemical Factors

SRB require both organic substrates and sulfate. In marine systems, with abundant supplies of sulfate, organic substrate concentration limits the rate of SR.[49] Depth profiles of SO_4^{2-} concentration in marine sediments have been successfully modeled using sulfate-independent diagenesis models as described in previous sections.[47,49]

The literature concerning the concentration below which sulfate becomes limiting to SR is rather ambiguous. Ramm and Bella[50] found a kinetic threshold for sulfate of 3000 μM. They concluded that at concentrations lower that 3000 μM sulfate would likely be limiting. Boudreau and Westrich[49] reported that SR rates in Long Island Sound sediments were independent of sulfate concentration at concentrations greater than 3 mM. Middleton and Lawrence[51] reported that growth of acetate utilizing SRB is independent of sulfate concentration at sulfate levels greater than 500 μM at 20°C and 100 μM at 31°C. Ingvorsen et al.,[52] in studying Lake Mendota sediments, found that SR was independent of sulfate concentration above 100 μM at 12°C. The half-saturation constant (K_m) for sulfate uptake in *Desulfobacter postgatei* was 200 μM.[53] Ingvorsen and Jørgensen[54] found K_m values ranging from 5 to 77 μM in four species of *Desulfovibrio*. Thus, it is not clear at exactly what concentration sulfate becomes limiting, and it is likely to vary somewhat depending on the dominant strain of SRB in any given habitat.

The average lake water concentration of sulfate is 110 μM.[55] Therefore, based on the above information, it is likely that in most freshwater lakes sulfate would be limiting to SRB. Studies in Wintergreen Lake[36] and Lake Anna[35] have shown that sulfate is depleted rapidly within the top 5 to 10 cm of sediment. Many studies have also demonstrated that the addition of sulfate to freshwater sediments results in increased rates of SR at the expense of methanogenesis.[17,56,57] Thus, in all but the most oligotrophic lakes, sulfate is the limiting factor for SR. Increases in lake water sulfate concentration associated with atmospheric sulfur deposition would increase the concentration of the limiting element resulting in an increase in alkalinity generation by SRB.

b. Physical Factors

Physical factors influence the rate of SR by controlling the onset and distribution of anoxic conditions and by controlling the rate of transport of sulfate and organic substrates to the anoxic areas where SR occurs. Schindler and Turner[5] concluded that the alkalinity generated by sulfate reduction in Lake 223 was a function of hypolimnetic volume, light penetration, and the concentrations of sulfate and oxygen. Larger anoxic hypolimnia will entrap more sulfate for reduction so that more alkalinity will be produced. The amount of mixing that occurs during lake turnover will determine the oxygen and sulfate concentrations in the hypolimnion. While less mixing means less oxygen in the hypolimnion and a quicker onset of SR, it also means that sulfate movement to the reducing zone will similarly be impeded. Schindler and Turner[5] found that during years with more complete mixing there was more sulfate reduced.

In many lakes, the majority of the SR occurs in the sediments. In these lakes, the rate of sulfate supply to the sediments controls the rate of SR. Typically sulfate enters the sediment via diffusion from the overlying lake water. Baker et al.[58] reported a mean k_{sulf}, a sulfate removal coefficient, of 0.46 (± 0.30) m/year for 14 softwater lakes. This coefficient can be thought of as a piston velocity relating how many meters of water per year must be processed by the sediments to account for the observed sulfate removal. They found that k_{sulf} was inversely related to the lake residence time. Lakes with a large residence time allowed more time for sulfate to diffuse into the sediment to be reduced and thus had a higher sulfate retention. Lakes with a short residence time had a low sulfate retention ($<10\%$ of input) since the sulfate is rapidly flushed out of the lake before coming into contact with the sediments.

In Lake Anna, calculated values of k_{sulf} were 12 to 14 m/yr, almost two orders of magnitude greater than the values reported by Baker et al.[58] The implications of this drastic difference could be attributed to different sulfate removal mechanisms. The lakes sampled by Baker et al.[58] were all affected by acid precipitation, not AMD, and diffusion was said to be the major mechanism for sulfate transport into the sediments. In Lake Anna diffusion could account for no more than 5% of the observed sulfate removal from the lake.[10] A mechanism with a more rapid sulfate transport rate could account for the higher k_{sulf} observed in Lake Anna. It has been hypothesized that sulfate is transported to the sediment via adsorption onto settling solid particles or iron floc. Therefore, the higher iron concentrations (10 to 50 ppm) in Lake Anna could account for the higher k_{sulf}.

Environmental temperature was found to be the major factor influencing SR rates in salt marsh sediments.[59] Abdollahi and Nedwell[60] showed that SR could be related to temperature by an Arrhenius function of the form,

$$\text{SR rate} = A * e^{(-E/RT)} \tag{36}$$

where A is the Arrhenius constant, E is the activation energy, R is the gas constant, and T is the temperature (Kelvin). In the Colne point salt marsh, Abdollahi and Nedwell[60] observed activation energies of 18.6 to 26.3 kcal/mol. However, Smith and Klug[36] observed no seasonality in sulfate turnover times in freshwater Wintergreen Lake. As a result, they hypothesized that changes in SR rates would be expected as a function of sulfate rather than temperature in a sulfate-limited freshwater system.

In the Contrary Creek arm of Lake Anna, sulfate transport to the sediment was the major factor limiting the rate of SR. Sulfate concentration profiles decrease rapidly with depth and are below the detection limit (10 μM) by 5- to 9-cm depth.[35]

Volatile fatty acids (acetate and propionate) have been detected in the sediments during all seasons of the year,[61] suggesting that organics do not limit SR in this system. Methanogenesis has been measured below the depth of sulfate depletion, indicating that suitable organic substrates exist, but that SR is absent at depth due to the lack of sulfate.

IV. EFFICIENCY OF SEDIMENT MICROBES IN NEUTRALIZING ACIDIC INPUTS

An important question related to the ability of microbial communities to neutralize acid pollution is "How fast can the microbes reestablish circumneutral conditions?" As seen above, Herlihy and Mills[22] demonstrated that pH conditions in the sediments underlying the acidified portion of Lake Anna were at about the same pH as the unacidified portion of the lake, i.e., pH ≈ 6.5. When Hurricane Danny swept through central Virginia in the summer of 1985, an opportunity was gained to determine how rapidly the anaerobic microbial communities of the Lake Anna sediments could reestablish the prestorm conditions. Danny was a highly erosive storm; about 11 cm of material was deposited on the sediment surface in the Contrary Creek arm. Samples were collected over the next 12 weeks and analyzed for a number of parameters related to both the anaerobic community and to the acid state of the sediment.[20] Over the period, the pH of the sediment increased from its initial value in the new sediment (<5) to the more typically observed circumneutral values in just a few weeks (Figure 3). Similarly, the Eh decreased and the alkalinity increased. SR in these sediments did not coincide with the initial increase in pH and alkalinity, which led Bell et al.[20] to conclude that some reaction other than SR accounted for the initial neutralization of the acidity.

In this case, the reduction of iron was most likely responsible for immediate neutralization of acidity in the fresh sediment, although some fermentation reactions and ammonification cannot be ruled out. Microbiological iron reduction would be expected to commence as soon as the oxygen was depleted in the sediment, and prior to SR. Because each mole of Fe^{3+} reduced consumes 3 mol of H^+ (see Equations 19 and 21), a substantial amount of acidity could be neutralized by this reaction in this instance.

Based on the Hurricane Danny study and additional evidence gathered since, it is obvious that iron plays a key role in the neutralization of acidity in Lake Anna, and likely plays a key role in many other situations as well.

V. THE ROLE OF IRON IN MICROBIOLOGICAL NEUTRALIZATION OF ACIDITY

In an earlier section, iron reduction was stated to be a potential reaction of importance in neutralization of acidity. In addition to direct acid neutralization, iron can also play a role in the delivery of SO_4^{2-} to anaerobic sites such as sediments, as well as influencing

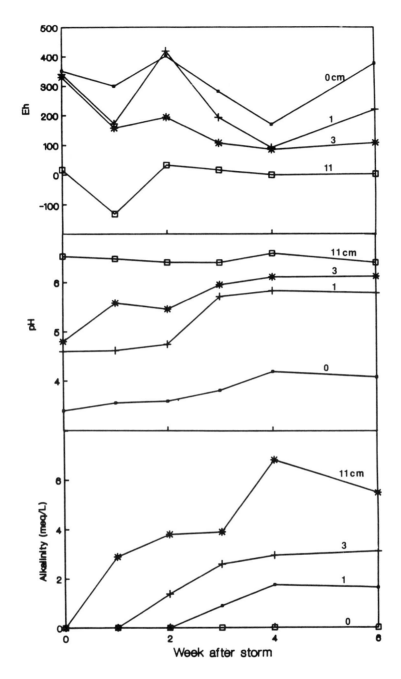

FIGURE 3. Sediment isodepths of Eh, pH, and alkalinity after a major storm event. The lines trace a physical property at a depth with time after the storm in the freshly deposited sediment (0, 1, and 3 cm) as well as in the "old" sediment material (11 cm).

the final products of other microbial processes such as SR. Figure 4 shows a conceptual model of these interactions of iron with sulfate and SR end products. Microcosm experiments utilizing sediment and water from Lake Anna have demonstrated that sulfate removal is greatly enhanced by the addition of Fe^{2+} to sulfate-rich water overlying reducing sediments.[43] The oxidation of the iron generates an iron oxyhydroxide floc that coprecipitates and adsorbs sulfate, causing enhanced delivery of SO_4^{2-} to the sediment. As pointed out by Herlihy et

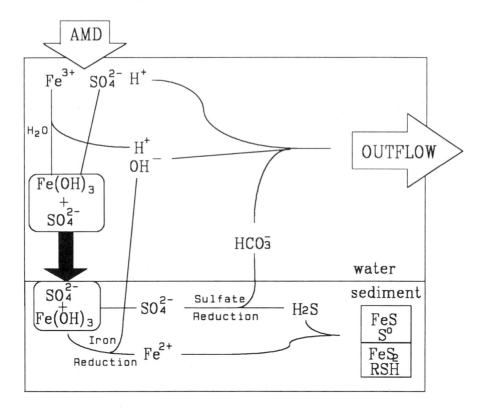

FIGURE 4.   Conceptual model of the interplay between iron and sulfur in the generation of alkalinity in Lake Anna water and sediments.

al.,[10] diffusion of sulfate into the sediments in Lake Anna accounted for only 5% of the observed sulfate retention in those sediments. Based on the microcosm experiments,[43] the additional sulfate retention could be explained by iron precipitation. Although there is acidity formed during the hydrolysis of iron leading to the formation of the precipitate, that acidity is neutralized in the subsequent step.

Upon reaching the sediment water interface, the iron oxyhydroxide dissolved as iron is reduced, both by direct bacterial activity and by redox reactions induced by microbial reactions not directly involving iron.[62] Reduction of iron neutralizes the acidity formed during the hydrolysis, and liberates adsorbed and precipitated SO_4^{2-} into the anaerobic sediments. At this point, both iron reduction and SR contribute to the neutralization of the acid mine drainage. Much of the neutralization contributed by the iron reduction goes to balance the acidity generated during the ferric iron hydrolysis. Any particulate ferric iron entering the lake can contribute to neutralization of the protons in the acid pollution. SR contributes alkalinity the neutralization of influent protons. The relative importance of the two reactions in alkalinity generation is not yet understood.

Iron also influences the overall neutralization of (sulfuric) acidity by precipitation of the SR end products. Although many workers have reported that the ultimate fate of SR end products is organic sulfur,[38,39] when iron is high, as in Lake Anna, most of the sulfur is held in iron monosulfides.[43] Furthermore, Baker and Brezonik[63] report that when iron was added to samples from Little Rock Lake in Wisconsin, the amount of iron monosulfide increased over unamended treatments.

In order to preserve the alkalinity generated by SR, the sulfides must be retained in the reduced form, or their oxidation will generate protons and consume the alkalinity. Precipitation of reduced metal sulfides serves as one mechanism for preservation of reduced sulfide as does liberation of H_2S. (The H_2S is released to the atmosphere and thus becomes unavailable for oxidation in the lake where it was reduced. Formation and release of $H_2S(g)$ is not a permanent sink for the sulfide, but it succeeds in moving the problem to another site.) Given that the reduction of iron plays some role in the neutralization of acidity, it can also be said that sulfides play a crucial role in trapping the reduced iron in the anaerobic sediments, preventing reoxidation and regeneration of the acidity.

Thus, iron is a critical element in the biogeochemical neutralization of acid pollution. In lakes where iron is not abundant, other elements may also play an important role, i.e., aluminum oxides might help remove sulfate from water as would any solid-phase anion exchange complexes; reduction of elements such as manganese can contribute to the alkalinity generation of some waters. While iron is not well studied with respect to reactions such as those described here, the role of the other elements is totally undocumented and also deserves attention to arrive at a complete picture of the overall neutralization process.

VI. SUMMARY

Microorganisms in all environments play critical roles in cycling of elements and mineralizing organic matter. That role is not abandoned in acidified environments, although different organisms may be involved and many of the processes may be lessened in terms of reaction rates. An additional role of microorganisms in acidic habitats is the generation of alkalinity by reduction of several of the components of the pollution itself. A complete understanding of the homeostatic neutralization process has not been reached, especially in terms of general concepts that can be applied to all systems, whether weakly or strongly acidified. Reactions other than SR are not well studied, and their contribution to neutralization and alkalinity generation needs further work. The quantitative role of additional elements (such as iron) in the enhancement of the transport of reducible materials to the sediment and the contribution to alkalinity generation by its own reduction is unknown, but may be very important in all but the most dilute lakes. The entire area of microbial contribution to the biogeochemistry of acid environments is still fertile ground for research, and should continue to provide a wealth of new information for several years to come.

REFERENCES

1. **Cowling, E. B.**, Acid precipitation in historical perspective, *Environ. Sci. Technol.*, 16, 110A, 1982.
2. **Tuttle, J. H., Randles, C. I., and Dugan, P. R.**, Activity of microorganisms in acid mine water. I. Influence of acid mine water on aerobic heterotrophs of a normal stream, *J. Bacteriol.*, 95, 1495, 1968.
3. **Cosby, B. J., Hornberger, G. M., Galloway, J. N., and Wright, R. F.**, Time scales of catchment acidification, *Environ. Sci. Technol.*, 19, 1144, 1985.
4. **Schindler, D. W., Wagemann, R., Cook, R. B., Ruszczyunski, T., and Prokopowich, J.**, Experimental acidification of lake 223, Experimental Lakes area: background data and the first three years of acidification, *Can. J. Fish. Aquat. Sci.*, 37, 342, 1980.
5. **Schindler, D. W. and Turner, M. A.**, Biological, chemical and physical responses of lakes to experimental acidification, *Water Air Soil Pollut.*, 18, 259, 1982.
6. **Kelly, C. A., Rudd, J. W. M., Cook, R. B., and Schindler, D. W.**, The potential importance of bacterial processes in regulating the rate of lake acidification, *Limnol. Oceanogr.*, 27, 868, 1982.
7. **Berner, R. A., Scott, M. R., and Thomlinson, C.**, Carbonate alkalinity in the porewaters of anoxic marine sediments, *Limnol. Oceanogr.*, 15, 544, 1970.
8. **Ben-Yaakov, S.**, pH Buffering of pore water of recent anoxic marine sediments, *Limnol. Oceanogr.*, 18, 86, 1973.
9. **Cook, R. B., Kelly, C. A., Schindler, D. W., and Turner, M. A.**, Mechanisms of hydrogen ion neutralization in an experimentally acidified lake, *Limnol. Oceanogr.*, 31, 134, 1986.

10. **Herlihy, A. T., Mills, A. L., Hornberger, G. M., and Bruckner, A. E.,** The importance of sediment sulfate reduction to the sulfate budget of an impoundment receiving acid mine drainage, *Water Resour. Res.*, 23, 287, 1987.

11. **Mills, A. L.,** Acid mine waste drainage: microbial impact on the recovery of soil and water ecosystems, in *Soil Reclamation Processes,* Tate, R. L. and Klein, D., Eds., Marcel Dekker, New York, 1985, 35.

12. **Stumm, W. and Morgan, J. J.,** *Aquatic Chemistry,* 2nd ed., John Wiley & Sons, New York, 1981.

13. **Carpenter, J., Odum, W. E., and Mills, A. L.,** Leaf litter decomposition in a reservoir affected by acid mine drainage, *Oikos,* 41, 165, 1983.

14. **Wassel, R. A. and Mills, A. L.,** Changes in water and sediment bacterial community structure in a lake receiving acid mine drainage, *Microb. Ecol.,* 9, 155, 1983.

15. **Mills, A. L. and Mallory, L.,** Community structure changes in epilithic bacterial communities stressed by acid mine drainage, *Microb. Ecol.,* 14, 219, 1987.

16. **Lovely, D. R., Dwyer, D. F., and Klug, M. J.,** Kinetic analysis of competition between sulfate reducers and methanogens for hydrogen in sediments, *Appl. Environ. Microbiol.,* 43, 1373, 1982.

17. **Lovley, D. R. and Klug, M. J.,** Sulfate reducers can outcompete methanogens at freshwater sulfate concentrations, *Appl. Environ. Microbiol.,* 45, 187, 1983.

18. **Laanbroek, H. J. and Veldkamp, H.,** Microbial interactions in sediment communities, *Phil. Trans. R. Soc. Lond. B,* 297, 533, 1982.

19. **Mah, R. A.,** Methanogenesis and methanogenic partnerships, *Phil. Trans. R. Soc. Lond. B.,* 297, 599, 1982.

20. **Bell, P. E., Herlihy, A. T., and Mills, A. L.,** Reestablishment of anaerobic sediment bacterial communities and conditions after storm deposition of acidic, aerobic sediments, *Appl. Environ. Microbiol.,* in review.

21. **Konopka, A., Gyure, R. A., Doemel, W., and Brooks, A.,** Microbial sulfate reduction in extremely acid lakes, WRRC Tech. Rep. 173, Purdue University, West Lafayette, IN, 1985.

22. **Herlihy, A. T. and Mills, A. L.,** The pH regime of sediments underlying acidified waters, *Biogeochemistry,* 2, 95, 1986.

23. **Gottschalk, G.,** *Bacterial Metabolism,* Springer-Verlag, New York, 1979.

24. **Schnoor, J. L. and Stumm, W.,** Acidification of aquatic and terrestrial systems, in *Chemical Processes in Lakes,* Stumm, W., Ed., John Wiley & Sons, New York, 1985, 311.

25. **Stone, A. T.,** Microbial metabolites and the reductive dissolution of manganese oxides: oxalate and pyruvate, *Geochim. Cosmochim. Acta,* 51, 919, 1987.

26. **Burdige, D. J. and Nealson, K. H.,** Microbial manganese reduction by enrichment cultures from coastal marine sediments, *Appl. Environ. Microbiol.,* 50, 491, 1985.

27. **Jones, J. G.,** Iron transformations by freshwater bacteria, in *Advances in Microbial Ecology,* Vol. 9, Marshall, K. C., Ed., Plenum Press, New York, 1986, 149.

28. **Goldhaber, M. B. and Kaplan, I. R.,** Controls and consequences of sulfate reduction rates in recent marine sediments, *Soil Sci.,* 119, 42, 1975.

29. **Postgate, J. R.,** *The Sulphate-Reducing Bacteria.,* Cambridge University Press, Cambridge, England, 1984.

30. **Jørgensen, B. B.,** A comparison of methods for the quantification of bacterial sulfate reduction in coastal marine sediments, *Geomicrobiol. J.,* 1, 29, 1978.

31. **Sorokin, Y. I.,** Experimental investigation of bacterial sulfate reduction in the Black Sea using S-35, *Mikrobiologiya,* 31, 329, 1962.

32. **Jørgensen, B. B.,** The sulfur cycle of a coastal marine sediment (Limfjorden, Denmark), *Limnol. Oceanogr.,* 22, 814, 1977.

33. **Nedwell, D. B. and Abram, J. W.,** Bacterial sulphate reduction in relation to sulphur geochemistry in two contrasting areas of saltmarsh sediment, *Estuar. Coast. Mar. Sci.,* 6, 341, 1978.

34. **Jørgensen, B. B.,** A comparison of methods for the quantification of bacterial sulfate reduction in coastal marine sediments. I. Measurements with radiotracer techniques, *Geomicrobiol. J.,* 1, 11, 1978.

35. **Herlihy, A. T. and Mills, A. L.,** Sulfate reduction in freshwater sediments receiving acid mine drainage, *Appl. Environ. Microbiol.,* 49, 179, 1985.

36. **Smith, R. L. and Klug, M. J.,** Reduction of sulfur compounds in the sediments of a eutrophic lake basin, *Appl. Environ. Microbiol.,* 41, 1230, 1981.

37. **Howarth, R. W. and Jørgensen, B. B.,** Formation of ^{35}S-labelled elemental sulfur and pyrite in coastal marine sediments (Limfjorden and Kysing Fjord, Denmark) during short-term $^{35}SO_4^-$ reduction measurements, *Geochim. Cosmochim. Acta,* 48, 1807, 1984.

38. **Rudd, J. W. M., Kelly, C. A., St. Louis, V., Hesslein, R. H., Furutani, A., and Holoka, M. H.,** Microbial consumption of nitric and sulfuric acids in acidified north temperate lakes, *Limnol. Oceanogr.,* 31, 1281, 1986.

39. **Nriagu, J. O. and Soon, Y. K.,** Distribution and isotopic composition of sulfur in lake sediments of northern Ontario, *Geochim. Cosmochim. Acta,* 49, 823, 1985.

40. **Zhabina, N. N. and Volkov, I. I.**, A method of determination of various sulfur compounds in sea sediments and rocks, in *Environmental Biogeochemistry and Geomicrobiology,* Vol. 3, Krumbein, W. E., Ed., Ann Arbor Science, Ann Arbor, MI, 1978, 735.

41. **Landers, D. H., David, M. B., and Mitchell, M. J.**, Analysis of organic and inorganic sulfur constituents in sediments, soils and water, *Int. J. Environ. Anal. Chem.,* 14, 245, 1983.

42. **Jørgensen, B. B., Fossing, H., and Thode-Andersen, S.**, Radiotracer studies of pyrite and elemental sulfur formation in coastal sediments, *EOS,* 65, 906, 1984.

43. **Herlihy, A. T.**, Sulfur Dynamics in an Impoundment Receiving Acid Mine Drainage, Ph.D. thesis, University of Virginia, Charlottesville, 1987.

44. **Mach, C. E. and Brezonik, P. L.**, Changes in trace metal concentrations in experimentally acidified Little Rock Lake, Vilas County, Wisconsin, Abstr. Ann. Meet. Am. Soc. Limnology and Oceanography, Madison, 1987, 49.

45. **Cook, R. L. and Schindler, D. W.**, The biogeochemistry of sulfur in an experimentally acidified lake, *Ecol. Bull. (Stockholm),* 35, 115, 1983.

46. **Berner, R. A.**, *Early Diagenesis: A Theoretical Approach,* Princeton University Press, Princeton, NJ, 1980, 149.

47. **Berner, R. A.**, An idealized model of dissolved sulfate distribution in recent sediments, *Geochim. Cosmochim. Acta,* 28, 1497. 1964.

48. **Jørgensen, B. B.**, A comparison of methods for the quantification of bacterial sulfate reduction in coastal marine sediments. II. Calculation from mathematical models, *Geomicrobiol. J.,* 1, 29, 1978.

49. **Boudreau, B. P. and Westrich, J. T.**, The dependence of bacterial sulfate reduction on sulfate concentration in marine sediments, *Geochim. Cosmochim. Acta,* 48, 2503, 1984.

50. **Ramm, A. E. and Bella, D. A.**, Sulfide production in anaerobic microcosms, *Limnol. Oceanogr.,* 19, 110, 1974.

51. **Middleton, A. C. and Lawrence, A. W.**, Kinetics of microbial sulfate reduction, *J. Water Pollut. Control. Fed.,* 1659, 1977.

52. **Ingvorsen, K., Zeikus, J. G., and Brock, T. D.**, Dynamics of sulfate reduction in a eutrophic lake, *Appl. Environ. Microbiol.,* 42, 1024, 1981.

53. **Ingvorsen, K., Zehnder, A. J. B., and Jørgensen, B. B.**, Kinetics of sulfate and acetate uptake by *Desulfobacter postgatei, Appl. Environ. Microbiol.,* 47, 403, 1984.

54. **Ingvorsen, K. and Jørgensen, B. B.**, Kinetics of sulfate uptake by freshwater and marine species of *Desulfovibrio, Arch. Microbiol.,* 139, 51, 1984.

55. **Wetzel, R. G.**, *Limnology,* 2nd ed., W. B. Saunders, Philadelphia, 1983.

56. **Lovely, D. R. and Klug, M. J.**, Intermediary metabolism of organic matter in the sediments of a eutrophic lake, *Appl. Environ. Microbiol.,* 43, 552, 1982.

57. **Winfrey, M. R. and Zeikus, J. G.**, Effect of sulfate on carbon and electron flow during microbial methanogenesis in freshwater sediments, *Appl. Environ. Microbiol.,* 33, 275, 1977.

58. **Baker, L. A., Brezonik, P. L., and Pollman, C. D.**, Model of internal alkalinity generation: sulfate retention component, *Water Air Soil Pollut.,* 31, 89, 1986.

59. **Nedwell, D. B. and Abram, J. W.**, Relative importance of temperature and electron donor and electron acceptor concentrations on bacterial sulfate reduction in salt marsh sediment, *Microb. Ecol.,* 5, 67, 1979.

60. **Abdollahi, H. and Nedwell, D. B.**, Seasonal temperature as a factor influencing bacterial sulfate reduction in a salt marsh sediment, *Microb. Ecol.,* 5, 73, 1979.

61. **Snyder, R. L. and Mills, A. L.**, Patterns of anaerobic carbon mineralization in an impoundment receiving acid mine drainage, unpublished data.

62. **Bell, P. E. and Mills, A. L.**, Biogeochemical conditions favoring formation of magnetite during anaerobic iron reduction, *Appl. Environ. Microbiol.,* 53, 2610, 1987.

63. **Baker, L. A. and Brezonik, P. L.**, personal communication.

Chapter 2

EFFECTS OF LAKE ACIDIFICATION ON MICROBIAL POPULATIONS AND PROCESSES

Salem S. Rao and B. Kent Burnison

TABLE OF CONTENTS

ABSTRACT

Bacteriological data collected for water and sediment cores from some Ontario lakes receiving acidic deposition indicate that bacterial populations and activities were diminished by 20 to 30% under acidic conditions. A pH value below 5.5 appeared to be critical for active populations. Measurements such as direct counts of total and respiring bacteria, heterotrophic plate counts, nitrifying and sulfur cycle bacteria, microbial activities (O_2 consumption rates and organic substrate utilization), and bacterial morphology and physiology were considered. An ultrastructural analysis is shown to differentiate between different levels of acid stress and combined acid/toxic metal stress under laboratory conditions. Data, methodology, and some implications of the studies are presented.

I. INTRODUCTION

During the last 2 decades it has become evident that precipitation over the eastern U.S. and Canada has become increasingly acidic.[1] The concern for the impacts of acidic deposition on aquatic ecosystems has led to several biogeochemical studies in Ontario and eastern Canada.[2] Some of the lakes near Sudbury, Ontario exhibit wide diversity in geochemical and biological characteristics and show varying degrees of stress from acid precipitation.[3-5] Acid rain fall (pH 4.0 to 5.0) has become a common event and one which is deleterious to many aquatic ecosystems.[6,7] In lakes, bacteria are an important component of the biota in terms of biomass and nutrient regeneration activity.[8] Bacterial activity can, in some cases, regulate the rate of lake acidification.[9] The effects of acidity on microbial activity in aquatic ecosystems have generated much interest recently.[9-15] In general, the field surveys show that bacterial populations and activity were lower in acidified lakes than in nonacidified lakes.[16] Following such observations some laboratory studies were made to ascertain the effects of acid stress on bacterial isolates[17,18] or obtained from artificially acidified water.[19] Few previous studies have related the bacterial activity to the *in situ* pH of the sediments.

The goal of this Chapter is to explore the nature and extent of the effects of acidic deposition on lake bacteria in some Ontario lakes (Figure 1) and the use of these populations from lake sediments as indicators of acid stress. An emphasis is placed on those parameters which correlate with the diminished capacity for degrading organic materials. The philosophy and ideas that form the basis for our studies here in Ontario may be applicable to developing studies on lakes receiving acidic deposition in Western Canada. The contents of this chapter are extracted from previously published scientific articles.[20-24]

II. MATERIALS AND METHODS

A. Study Site

The lakes examined are located within a radius of 30 km from the smelters in Sudbury, Ontario (Figure 1). These Precambrian shield lakes encompass a wide diversity of physical and chemical characteristics and show varying degrees of stress from acid rain.[24] Acidity of the lakes can be traced back to acid rain. The details of the geological settings and the general limnology of lakes in the region have been outlined.[5]

B. Sample Collection

Surface water samples (1 l) from a series of eight lakes (Figure 1) were collected in sterile containers. These samples were packed in ice and transported within 48 h to our laboratory for further processing. This procedure did not affect the levels of bacterial populations.[20] The studies were usually carried out during May through October. Sediment cores (50 × 6 cm) were collected by means of a lightweight coring device.[16] Only cores that came up with their sediment-water interfaces intact and undeformed were subsectioned and processed

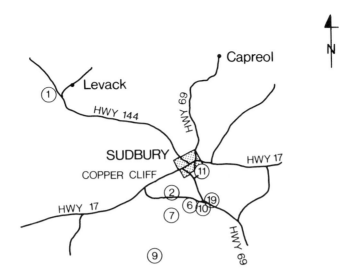

FIGURE 1. Microbiology sampling sites in lakes receiving acid precipitation near Sudbury, Ontario. 1, Windy Lake; 2, Hannah Lake; 6, Silver Lake; 7, Lohi Lake; 9, Wavy Lake; 10, McFarlane Lake; 11, Ramsey Lake; 19, Richard Lake; and 20, Clearwater Lake. (From Rao, S. S., Jurkovic, A. A., and Nriagu, J. O., *Environ. Pollut. Ser. A*, 36, 195, 1984. With permission.)

within 1 h after retrieval. Each core was subsectioned at 1 to 2 cm intervals up to 40 cm. The samples were collected in sterile plastic bags and refrigerated immediately. The samples were processed within 48 h in the shore laboratory.

The pH measurements were made using a portable pH meter with the electrodes inserted directly into the mud. Organic content of the sediments was measured by combustion [loss on ignition (LOI)] according to standard procedure.[27]

III. BACTERIOLOGICAL PROCEDURE

Counting of bacteria was performed using epifluorescence microscopy after staining with acridine orange. The staining of filters and solutions used follow Zimmerman et al.[28]

Total and actively respiring bacterial populations were estimated microscopically on all water samples using the combined acridine orange INT-formazan reduction technique.[28] Reduction of the tetrazolium dye in bacterial cells involves respiration. In actively respiring bacteria, deposits of accumulated reduced INT-formazan, seen as optically dense dark-red intracellular spots, can be seen by light microscope observation.[28] Total bacteria were estimated on the same slide using UV light. Numbers of bacteria were determined in 20 ocular fields, resulting in a total count of 30 to 300 bacteria.

Aerobic heterotrophic bacteria were measured in all samples using the spread plate procedure and a low-nutrient medium. Incubation of the inoculated plates was for 7 d at 20°C.[29]

IV. OXYGEN CONSUMPTION RATE

Oxygen consumption by the various sediment fractions from the lakes was measured at 20°C using a Gilson differential respirometer. Each Warburg flask contained 3 g of sediment (wet weight), 2 ml of filter-sterilized lake water, which served as diluent, and 0.2 ml of 20% KOH in the center well for carbon dioxide absorption. A preincubation period of 15 min was used in all experiments. No exogenous substrate was added. The Warburg flasks were shaken at a rate of 100 strokes per minute to ensure that the oxygen diffusion rate was not a limiting factor. KCN was used to stop microbial activity. By calculating the difference between total oxygen uptake and oxyten uptake obtained by the KCN poisoning, microbial oxygen uptake was determined.

V. HETEROTROPHIC ACTIVITY METHOD

Heterotrophic activity measurements were made using modifications of Harrison et al.[30] Sediment (0.5 ml) was suspended with 10 ml of filter (0.2 μm)-sterilized lake water. Various aliquots of a 18.5 kBg/ml solution of uniformly labeled ^{14}C-glucose or ^{14}C-glutamic acid (Amersham, specific activity 10 and 10.4 GBg/mol, respectively) were added to combusted (450°C for 2 h) glass serum bottles. All concentrations were run in duplicate. Then 5 ml of filter sterilized lake water was added. At zero time, 0.5 ml of the diluted sediment slurry was added to each bottle. Appropriate killed controls consisted of adding 100 μl of form-aldehyde before the addition of sediment. The respiration cup assembly, with a 2.5×5.0 cm accordion-folded piece of filter paper (Whatman No. 1) in the cup, was used as the bottle stopper. The bottles were incubated in the dark for 1 h at 20°C. At the end of the incubation, 200 μl of 5 N H_2SO_4 was added to each bottle to stop further isotope uptake and release the $^{14}CO_2$ from the aqueous phase. β-Phenethylamine (200 μl) was carefully added to the cup assembly. The CO_2 was allowed to evolve for 2 h, the bottles were uncapped, and the filter paper added to a scintillation vial containing 10 ml of ACS II (Amersham). The vials were counted on a liquid scintillation counter using an external standard for quench correction.

The suspended sediments were collected on 0.45 μm Sartorius membrane filters (25 mm). Distilled water (10 ml) was used to wash the residue from the bottles, rinse the filter funnels, and rim the edges of the filters after the funnel had been removed. The filters were placed in glass scintillation vials, 10 ml ACS II added, and the filters were allowed to dissolve. The sediment was dispersed with a Sonic Dismembrator (Model 150, Artek Systems Corp.) at the full power setting. Next, 5 ml of water was added to the sediment suspension, mixed vigorously, and counted as above.[23]

VI. ORGANIC SUBSTRATE UTILIZATION

Indigenous bacterial populations taken directly as inoculum from the two lakes at the extremities of our selected pH range (Silver Lake, about pH 4.0 and McFarlane Lake about pH 7.0) were harvested from 48 h batch cultures (at room temperature) in one-half strength nutrient broth at pH 4.0 and pH 7.2. Cells were washed three times with filter-sterilized, low-response water and harvested by centrifugation at 10×1000 g for 20 min at 4°C. These washed cells were resuspended (40 ml) for the organic substrate utilization studies. Bacterial respiration using three substrates (glucose, glutamic acid, and sodium acetate) at 5 μmol of substrate per flask was determined using a Gilson differential respirometer at 20°C. Respiration studies were performed in duplicate. Data were corrected for endogenous respiration. Respiration was used to infer the bacterial activity index.

VII. MORPHOLOGICAL STUDIES

The chemical preparation of the samples for electron microscopy and the analysis of sections was done according to Leppard et al.,[31] but with one change; the phosphate buffer was replaced by cacodylate-HCl buffer. Counterstaining of epoxy-embedded samples[32] and the mode of analysis of ultrathin sections (50 to 70 nm) was done according to Burnison and Leppard.[33] Observations and photographs were made with a Philips 300 transmission electron microscope (TEM).

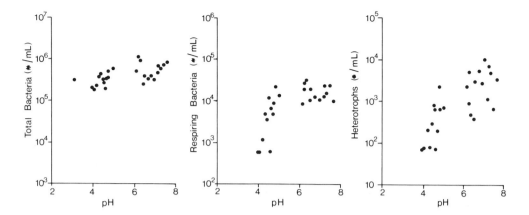

FIGURE 2. Relationships between bacteria and pH in lake water receiving acid precipitation near Sudbury, Ontario, 1982. (From Rao, S. S., Jurkovic, A. A., and Nriagu, J. O., *Environ. Pollut. Ser. A*, 36, 195, 1984. With permission.)

VIII. RESULTS AND DISCUSSION

Figure 2 is a scattergram showing the representative relationship between bacterial populations and pH in waters of the eight lakes studied. The pH values of these lakes did not change significantly during the study period of May to October. Total bacterial densities were in the 10^6 ml^{-1} range in all lakes examined. Acid stress had no apparent effect on total bacteria since lakes having a pH of 7.8 had total bacterial populations somewhat similar to lakes with pH 3.8. These data confirm earlier observations,[12,34] which showed no consistent relationship between total bacteria counts and water acidity.

Respiring bacteria and aerobic heterotrophs in contrast showed a definite response to environmental pH changes. Generally, maximum respiring bacteria recorded in these lakes were in the 10^5 ml^{-1} range while aerobic heterotrophs recorded were in the 10^4 ml^{-1} range. The pH values below about 5.5 may be critical for these populations.

The aerobic heterotrophic count dropped significantly around pH 5.5. The low-nutrient media used a pH of 7.0, but we assumed that most bacteria would grow well at this pH. Boylen et al.[34] isolated bacteria from lakes of various pH values grown on media at pH 7.0 and 5.0 and found that all colonies grew best on the media of higher pH, irrespective of the pH of the lake from which they were collected. Bacteria present in our lakes which are at a pH lower than 5.0 may not necessarily grow at pH 7.0. The lower count which is observed at pH 4 (Figure 2) must be viewed with this reservation.

The respiring bacteria densities also seem to decline around pH 5.5. Visual observations of the INT reduction slides indicate that the bacteria present in lakes with a low pH are smaller than those from a circumneutral pH. The observed decline may be caused by an actual reduction of metabolic activity of bacteria at low pH or possibly there is a threshold visual limitation of observing a formazan structure inside a very small bacterial cell.

Figure 3 compares the pH profiles with changes in aerobic heterotrophs in the lake sediment cores from the acid-stressed Silver Lake and the nonacid-stressed McFarlane Lake. The aerobic heterotrophic bacterial populations in the top 5-cm layer of sediment cores from Silver Lake (pH of 3.8 to 5.0) number in the 10^3 to 10^4 ml^{-1} range. However, their densities between 5 and 40 cm, where the recorded pH was about 5.5, were relatively larger (10^5 to 10^6 ml^{-1}). In contrast, the profile of these bacterial populations in the core from nonacid McFarlane Lake declined from about 10^7 ml^{-1} in the surficial sediments with a pH of about 7.1 to about 10^6 ml^{-1} in the deeper layers, where the pH is relatively constant at about 6.5. These data suggest that pH below about 5.5 appears to be critical for bacterial populations in the sediments.

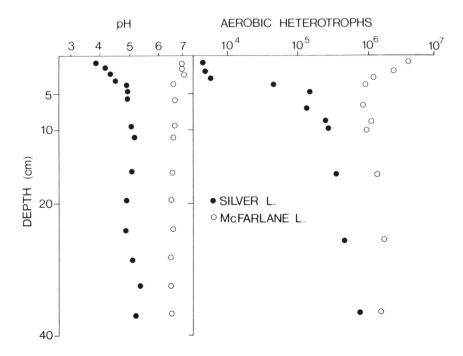

FIGURE 3.   Effects of acid stress on bacterial populations in lake sediments, (From Rao, S. S., Jurkovic, A. A., and Nriagu, J. O., *Environ. Pollut. Ser. A,* 36, 195, 1984. With permission.)

A. Morphological Effects of Low pH on Bacteria

A dominant mixed population of bacteria was isolated from nonacid-stressed McFarlane Lake and subjected to low-pH stress (pH range of 3 to 7). At pH level of 3, 4, 5, 6, and 7, the culture was repeatedly subcultured in order to permit its adjustment to the new pH environment. Growth was too sparse at pH 3.0 for proper microscopic analysis. Lowering of the pH of the aquatic environment seems to increase the complexity of the cell wall-environment interface and the abundance of extracellular materials.[22]

Ultrastructural analysis has also been used to differentiate between different levels of both acid stress and combined acid/copper stress in laboratory experiments on bacterial enrichment cultures).[24]

B. Sediment Microbial Respiration and Bacterial Organic Substrate Utilization

One of the effects of lake acidification is reported to be the increased accumulation of organic matter in the lake sediments.[14,17,18,35] Kelly et al.[36] have shown that decomposition rates of "old" organic carbon was unaffected by pH values as low as 4.0, but newly sedimented material decomposition rates started to decrease at pH 5.25 to 5.0. Figure 4 compares the concentration of organic matter in the sediments with the pH of the overlying water for the eight lakes studied. The strong relationship found confirms the results of the previous studies. The build-up of organic matter has been attributed to the inhibition of microbial decomposition processes and/or to a shift from bacterial to less efficient fungal mineralization.[14,17,18,35] The increased storage of organic matter in the sediments is tied to the recycling of nutrients and hence can affect the behavior of the whole lake ecosystem. This particular facet of lake acidification remains to be investigated in detail.

Figure 5 shows that a strong relationship exists between low-pH stress and microbial activity. For example, as the pH decreased, a corresponding decrease in the rate and extent of oxygen utilization and bacterial organic substrate utilization was observed. Maximum uptake at the end of 3 h of incubation in the top 5 cm layer of acid-stressed Silver Lake did

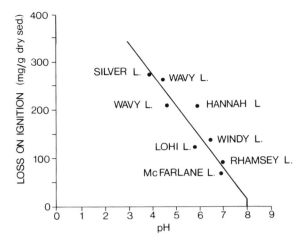

FIGURE 4. pH and total organics in lake sediments. (From Rao, S. S., Jurkovic, A. A., and Nriagu, J. O., *Environ. Pollut. Ser. A,* 36, 195, 1984. With permission.)

not exceed 10 to 80 μl O_2 per 10^9 cells per milliliter. However, for the same incubation period in the top 5 cm layer of nonacid-stressed McFarlane Lake, oxygen uptake was nearly ten times more. Somewhat similar observations (3 to 30 times increase) were noticed with regard to indigenous bacterial respiration (organic substrate utilization) from the nonacid McFarlane Lake. Using the INT-formazan reduction technique, at pH 4 no respiring bacteria were detected at the end of 3 h using any of the three organic substrates. However, under the same conditions, the densities of respiring bacteria were in the range of 10^9/ml when the pH was 7.2. No actively respiring bacteria were detected under severe acid-stress conditions. However, this does not necessarily mean such populations were absent; possibly the technique applied did not permit detection of such populations.[22]

The heterotrophic activity method of Wright and Hobbie[37] was used to estimate the level of heterotrophy in the sediments from the three Ontario lakes.[23] We have assumed that all the activity was from bacterial origin since we used low levels of substrate. The limitations of the heterotrophic activity procedure were found by Wright and Burnison.[38] The most important drawback, under the conditions in which it has been used here, is that only one organic compound can be tested at a time and this does not reflect the actual flux of the entire dissolved organic carbon pool, which is a mixture of organic compounds.

Figure 6 illustrates calculated uptake velocities (V_{50}) for ^{14}C-glucose and ^{14}C-glutamic acid in the sediments of the three Ontario lakes studied. Usually glucose uptake is faster than glutamic acid uptake with the exception of Clearwater Lake (water column pH 4.5) sediment. Two possibilities for this observation are (1) the bioavailability of the ^{14}C-labeled compound is higher in Clearwater Lake sediment than in the other two sediments and (2) there is an acidophilic bacterial population which is more efficient at taking up glutamic acid. Although the second possibility cannot be completely ignored, we feel that glutamic acid adsorption to calcareous minerals or clay in the circumneutral McFarlane Lake is the most likely explanation. Precise research on the proper adsorption control is still needed.[23]

The various observations on the effects of acid precipitation on the microbial population and its activity in lake sediments are summarized in a conceptual model (Figure 7). Lake acidification is surmised to reduce the bacterial activity and hence increase the organic content of the sediments. These parameters can, therefore, be used to trace the historical changes in the response of lakes to acid precipitation.[22]

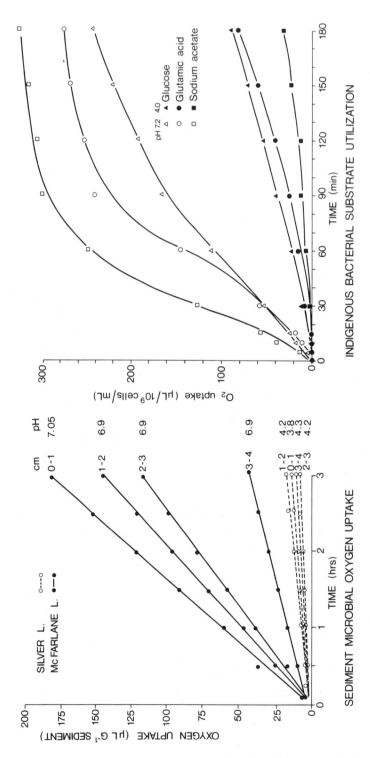

FIGURE 5.   Microbial activity from acid-stressed and nonacid-stressed lakes. (From Rao, S. S., Paolini, D., and Leppard, G. G., *Hydrobiologia*, 114, 115, 1984. With permission.)

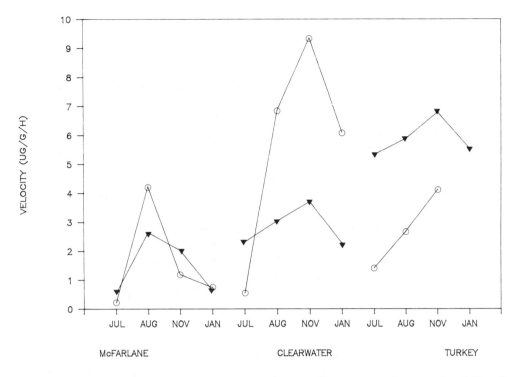

FIGURE 6.   Calculated velocities for glucose (▼) and glutamate (○) at an assumed concentration of 50 μg/l. (From Burnison, B. K., Rao, S. S., Jurkovic, A. A., and Nuttley, D. J., *Water Pollut. Res. J. Can.*, 21, 560, 1986. With permission.)

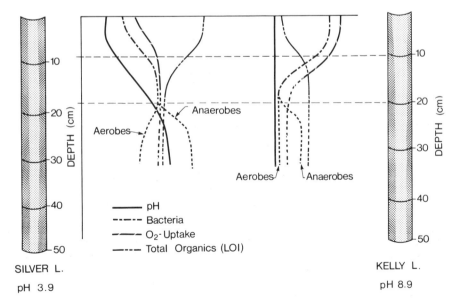

FIGURE 7.   Influence of acid precipitation on bacterial activity in lake sediment (conceptual model). (From Rao, S. S., Paolini, D., and Leppard, G. G., *Hydrobiologia*, 114, 115, 1984. With permission.)

IX. CONCLUSIONS

The following are the major conclusions from our studies:

1. Bacterial populations and densities were lower in acid-stressed lakes than in nonacid-stressed lakes.
2. Respiring and heterotrophic bacterial populations in acid-stressed lakes were nearly an order of magnitude less than in the nonacidified lake.
3. A pH value less than 5.5 may be critical for these bacterial populations in sediments and in the water column.
4. A relationship was found between low pH stress and sediment microbial activity.
5. Diminished microbial activity in surface sediments resulted in an increased accumulation of organic matter.
6. The increased storage of organic matter in the sediments is tied into the recycling of nutrients and hence can affect the behavior of the whole lake ecosystem.
7. The greatest complexity in terms of bacterial cell structure diversity and development of extracellular products is found at pH 5.0.
8. Low pH stress has a marked effect on the bacterial morphology per se and/or the selection of dominant types.
9. Low bacterial activity in acidified lakes may be the result of certain cell structural changes.
10. The physiological alterations in bacteria may be significant as stress indicates if related to cell surface exchange and/or membrane permeability properties associated with nutrient transport or transport of toxic substances.

REFERENCES

1. **Almer, B., Dickson, W., Ekstrom, C., and Hornstorm, E.,** Sulfur pollution and aquatic ecosystem, in *Sulfur in the Environment, Part II,* Nriagu, J. O., Ed., John Wiley & Sons, New York, 1978, 273.
2. **Nriagu, J. O.,** *Environmental Impacts of Smelters,* Nriagu, J. O., Ed., John Wiley & Sons, New York, 1984.
3. **Gorham, E. and Gordon, A. G.,** The influence of smelter fumes upon the chemical composition of lake waters near Sudbury, Ontario and upon the surrounding vegetation, *Can. J. Bot.,* 38, 477, 1960.
4. **Beamish, R. J. and Harvey, H. H.,** Acidification of lakes in Canada by acid precipitation and the resulting effects on fishes, *J. Fish. Res. Board Can.,* 29, 1131, 1972.
5. **Conroy, N. I., Hawly, K., and Keller, K.,** Extensive Monitoring of Lakes in the Greater Sudbury Area, Tech. Rep., Ontario Ministry of the Environment, Sudbury, 1978.
6. **Haines, T. A.,** Acidic precipitation and its consequences for aquatic ecosystems: a review, *Trans. Am. Fish. Soc.,* 110, 669, 1981.
7. **Wright, R. F. and Gjessing, E. T.,** Changes in the chemical composition of lakes, *Ambio,* 5, 219, 1976.
8. **Wetzel, R. G.,** *Limnology,* Wetzel, R. G., Ed., W. B. Saunders, Philadelphia, 1975, 571.
9. **Kelly, C. A., Rudd, J. W. M., Cook, R. B., and Schindler, D. W.,** The potential importance of bacterial processes in regulating rate of lake acidification, *Limnol. Oceanogr.,* 27, 868, 1982.
10. **Leivestad, H., Hendrey, G., Muniz, I. P., and Shekvik, E.,** Effects of acid precipitation on freshwater organisms, in *Impact of Acid Precipitation on Forest and Freshwater Ecosystems in Norway,* Braekke, F. N., Ed., SNSF Project F.R 6/76, Oslo, 1976, 87.
11. **Alexander, M.,** Effects of acidity on microorganisms and microbial processes in soil, in *Effects of Acid Precipitation on Terrestrial Ecosystem,* Hutchinson, T. C. and Haves, M., Eds., Plenum Press, New York, (published in coordination with NATO Scientific Affairs Division), 1978, 363.
12. **Traan, T. S.,** Effects of acidity on decomposition of organic matter in aquatic environments, in *Proc. Int. Conf. Ecological Impact of Acid Precipitation,* Drablos, D. and Tollan, A., Eds., SNSF Project, Oslo-Aas, Norway, 1980, 340.
13. **Pagel, J. E.,** A Review of Microbiological Processes Relevant to the Effects of Acid Precipitation, Tech. Rep., Ontario Ministry of the Environment, Sudbury, June, 1981.

14. **Fjerdingstad, E. and Nilssen, J. P.,** Bacteriological and hydrological studies on acidic lakes in Southern Norway, *Arch. Hydrobiol.*, 4, 443, 1982.
15. **Leduc, L. G. and Ferroni, G. D.,** Glucose mineralization activity and the use of the heterotrophic activity method in an acidified lake, *Water Res.*, 18, 609, 1984.
16. **Babich, H. and Stotzky, G.,** Influence of pH on inhibition of bacteria, fungi and coliphages by bisulfite and sulfite, *Environ. Res.*, 15, 405, 1978.
17. **Baker, M. D., Inniss, W. E., and Mayfield, C. I.,** Effect of pH on the growth and activity of heterotrophic sediment microorganisms, *Chemosphere*, 11, 973, 1982.
18. **Baker, M. D., Inniss, W. E., Mayfield, C. I., and Wong, P. T. S.,** Effect of acidification, metals and metalloids on sediment microorganisms, *Water Res.*, 17, 925, 1983.
19. **Baath, E., Lundgren, B., and Soderstron, B.,** Effects of artificial acid rain on microbial activity and biomass, *Bull. Environ. Contam. Toxicol.*, 23, 737, 1979.
20. **Rao, S. S. and Dutka, B. J.,** Influence of acid precipitation on bacterial populations in lakes, *Hydrobiologia*, 98, 153, 1983.
21. **Rao, S. S., Jurkovic, A. A., and Nriagu, J. O.,** Bacterial activity in sediments of lakes receiving acid precipitation, *Environ. Pollut. Ser. A*, 36, 195, 1984.
22. **Rao, S. S., Paolini, D., and Leppard, G. G.,** Effects of low-pH stress on the morphology and activity of bacteria from lakes receiving acid precipitation, *Hydrobiologia*, 114, 115, 1984.
23. **Burnison, B. K., Rao, S. S., Jurkovic, A. A., and Nuttley, D. J.,** Sediment microbial activity in acidic and non-acidic lakes, *Water Pollut. Res. J. Can.*, 21, 560, 1986.
24. **Leppard, G. G. and Rao, S. S.,** Acid stress and lake bacteria: ultrastructural and physiological correlates of some acid/copper stresses, *Hydrobiologia*, 160, 241, 1988.
25. **Nriagu, J. O., Wong, H. K. T., and Coker, R. D.,** Deposition and chemistry of pollutant metals in lakes around the smelters at Sudbury, Ontario, *Environ. Sci. Technol.*, 16, 551, 1982.
26. **Williams, J. D. and Pashley, A. E.,** Light-weight corer designed for sampling very soft sediments, *J. Fish. Res. Board Can.*, 36, 241, 1979.
27. **American Public Health Association,** *Standard Methods for the Examination of Water, Wastewater and Sediments*, 15th ed., APHA-AWWA-WPCF, Washington, D.C., 1980, 1134.
28. **Zimmerman, R., Iturriaga, R., and Brick, J. B.,** Simultaneous determination of the total number of aquatic bacteria and the number thereof involved in respiration, *Environ. Microbiol.*, 36, 926, 1978.
29. **Dutka, B. J.,** Methods for microbiological analysis of water, waste water and sediments, Inland Water Directorate, Burlington, Ontario, 1978.
30. **Harrison, M. J., Wright, R. T., and Morita, R. Y.,** Method for measuring mineralization in lake sediments, *Appl. Microbiol.*, 21, 698, 1971.
31. **Leppard, G. G., Massalski, A., and Lean, D. R. S.,** Electron-opaque microscopic fibrils in lakes: their demonstration, their biological derivation and their potential significance in the redistribution of cations, *Protoplasma*, 92, 289, 1977.
32. **Spurr, A. R.,** A low-density epoxy resin embedding medium for electron microscopy, *J. Ultrast. Res.*, 26, 31, 1969.
33. **Burnison, B. K. and Leppard, G. G.,** Isolation of colloidal fibrils from lake water by physical separation techniques, *Can. J. Fish. Aquat. Sci.*, 40, 373, 1983.
34. **Boylen, C. W., Shick, M. O., Roberts, D. A., and Singer, R.,** Microbiological survey of Adirondack lakes with various pH values, *Appl. Environ. Microbiol.*, 45, 1538, 1983.
35. **Grahn, O., Hultberg, H., and Landner, L.,** Oligotrophication — a self-accelerating process in lakes subjected to excessive supply of acid substances, *Ambio*, 3, 93, 1974.
36. **Kelly, C. A., Rudd, J. W. M., Furutani, A., and Schindler, D. W.,** Effects of lake acidification on rates of organic matter decomposition in sediments, *Limnol. Oceanogr.*, 29, 687, 1984.
37. **Wright, R. T. and Hobbie, J. E.,** Use of glucose and acetate by bacteria and algae in aquatic ecosystems, *Ecology*, 47, 447, 1966.
38. **Wright, R. T. and Burnison, B. K.,** Heterotrophic activity measured with radiolabelled organic substrates, in *Native Aquatic Bacteria: Enumeration, Activity, and Ecology*, Costerton, J. W. and Colwell, R. R., Eds., ASTM STP 695, 1979, 140.

Chapter 3

BIOGEOCHEMICAL CYCLING OF ORGANIC MATTER IN ACIDIC ENVIRONMENTS: ARE MICROBIAL DEGRADATIVE PROCESSES ADAPTED TO LOW pH?

Ronald Benner, David L. Lewis, and Robert E. Hodson

TABLE OF CONTENTS

I. INTRODUCTION

Microorganisms are the major agents responsible for the biogeochemical cycling of nutrients that often limit overall productivity of ecosystems and, in part, determine ecosystem structure. Many biochemical reactions involved in the carbon, nitrogen, phosphorus, and sulfur cycles, as well as in trace element cycles, are mediated exclusively or to large extent by bacteria.[1] In addition to their role in nutrient cycling, microorganisms are now recognized as important components of food webs in many environments.[2,3] Therefore, in studies of the effects of environmental stress on ecosystem structure and function, microorganisms and the processes they mediate should be of primary interest to ecologists. Microorganisms, however, can proliferate under environmental conditions that are often too stressful to support other forms of life. If microbial populations are capable of adapting to conditions of environmental stress, then microbial abundance and activity may not be significantly altered over the long term, and perturbations to other groups of organisms may dictate ecosystem-level responses to stress.

Comparisons of microbial activities between circumneutral and naturally acidic environments can provide insights into the long-term adaptability of microbial populations to acid stress, and thus may indicate which microbially mediated processes, if any, are sensitive to low pH conditions. In performing comparisons of this nature, we observed that the rates of microbial degradation of vascular plant detritus were severalfold lower in waters of the naturally acidic (pH 4) Okefenokee Swamp relative to microbial degradation rates in circumneutral pH waters from nearby marshes.[4] In particular, the polysaccharide components of plant material, which comprise 50 to 80% of the biomass of most vascular plants, were particularly resistant to degradation at low pHs. Rates of degradation of the lignin component of plant material, which comprises 5 to 10% of the biomass of herbaceous plant tissues, were less affected over the pH range from 4 to 8. Other measurements of microbial activity and abundance indicated that microbial biomass and rates of utilization of certain dissolved organic substrates in Okefenokee water and sediment were comparable to microbial biomass and activity in circumneutral pH wetlands.[5,6] In summary, these data suggested that although certain microbial degradative processes were inhibited by low pH, microbial biomass and activity were quite high in the acidic water and sediment of the Okefenokee Swamp.

The pH in Okefenokee water and sediment is low due to high concentrations of organic acids that leach from the abundant accumulations of decaying plant material. Thus, organic anions predominate in Okefenokee water rather than inorganic ions, such as sulfate and nitrate, which predominate in anthropogenically acidified environments. This distinction has been shown to be very important because organic acids are less toxic to many biological communities than mineral acids, presumably due to the capacity of organic acids to complex with and thus reduce concentrations of toxic metals such as aluminum.[7,8] Organic acids, such as tannins, also are known to complex with proteins and thereby inactive microbial enzymes involved in the degradation of plant material.[9,10] Therefore, we considered that the depressed rates of polysaccharide degradation that we observed in Okefenokee water might be due to some factor other than pH per se. Samples of Okefenokee water were charcoal-filtered to remove the organic acids and reacidified with mineral acid to pH 4.[4] Rates of polysaccharide degradation in this water were similar to rates in natural Okefenokee water, indicating that pH was the major factor causing the depressed rates of polysaccharide degradation.

In the present study, we compared the rates of microbial degradation of a variety of dissolved and particulate substrates in water and sediment from the Okefenokee Swamp, in southern Georgia, and Corkscrew Swamp, a circumneutral pH wetland in southern Florida. These two wetland ecosystems share many of the same types of plant communities and both are peat-forming systems. As in many wetlands, streams, and small lakes, vascular plant detritus is a major source of organic matter in these two freshwater swamps. The relationship between pH and rates of biodegradation of organic substrates was determined for natural microbial assemblages and for several bacterial isolates from these environments. Results from these studies suggest that microbial degradative processes that rely on extracellular enzymes are depressed at low pHs, whereas the microbial utilization of low molecular weight compounds that can be directly transported into cells is not substantially affected by variations in pH from 4 to 8. Furthermore, we suggest that microbial populations will not "adapt" for the rapid utilization of lignocellulosic substrates at low pHs.

II. MATERIALS AND METHODS

A. Sampling Sites

The Okefenokee Swamp is a large (1754 km^2) peat-forming wetland located in southern Georgia. Waters in the swamp range in pH from 3.4 to 4.2 and contain high concentrations of organic acids. Mizell Prairie is a marsh area in the swamp that is dominated by the sedge, *Carex walteriana*. Corkscrew Swamp Sanctuary is a relatively undisturbed portion of the Big Cypress Swamp that is owned by the National Audubon Society and is located in southwest Florida. Like the Okefenokee Swamp, Corkscrew Swamp is a peat-forming wetland; however, the pH of Corkscrew water ranges from 6 to 8. Central Marsh is a large area between cypress strands that is dominated by the sedge, *Mariscus jamaicensis*. All water and sediment samples collected in these two swamps were returned to our laboratories in Athens, GA, and used in incubations within 72 h of collection.

B. Preparation of Radiolabeled Lignocelluloses

The procedures used to prepare specifically radiolabeled [^{14}C-polysaccharide]-lignocellulose and [^{14}C-lignin]lignocellulose have been previously described.[11] Cuttings of the aboveground parts of C. walteriana were incubated under natural light for several days in sterile water containing either [^{14}C]glucose to label the polysaccharide components or [^{14}C]cinnamic acid to label the lignin component. Uniformly [^{14}C]-labeled lignocellulose was prepared by growing potted plants in an atmosphere containing $^{14}CO_2$ as previously described.[12] The labeled plants were then dried (50°C) and ground to pass a 40-mesh sieve (425 μm). Extractive-free lignocellulose was isolated from unincorporated radioactivity and other plant components by serial extraction of ground plant material in boiling ethanol, ethanol-benzene, and water.[11] Specific activities of the labeled lignocelluloses were determined by combusting weighed portions to [^{14}C]carbon dioxide in a biological oxidizer (OX-300, Harvey Instrument Co.) and radioassaying the trapped [^{14}C]carbon dioxide by liquid scintillation spectrometry. Chemical analyses of these preparations indicate that they are substantially free of radiolabeled contaminants.[11]

C. Determination of Bacterial Abundance

Bacterial abundance was determined by the acridine orange direct count (AODC) method.[13] Water samples, preserved with formalin, were stained with acridine orange (final concentration, 0.01%) and filtered through 0.2-μm pore-size Nuclepore filters that had been previously treated with Irgalan Black. Sediment samples were diluted with filter-sterilized water and sonicated for 1.5 min before filtering. At least 10 randomly selected microscope fields with ≥35 cells per field were counted on each filter with an Olympus BHS microscope.

D. Isolation and Identification of Bacterial Cultures

Bacteria were isolated using Standard Methods agar (BBL®) diluted 1:10 (vol/vol) with distilled water. Okefenokee bacteria were isolated and maintained at pH 4, and Corkscrew bacteria were isolated and maintained at pH 7. Species were identified through standard biochemical tests and morphological descriptions by Teska et al.[14]

Table 1

**BACTERIAL ABUNDANCE IN WATER AND
SEDIMENT FROM TWO FRESHWATER MARSHES
OF DIFFERENT pH: MIZELL PRAIRIE IN THE
OKEFENOKEE SWAMP AND CENTRAL MARSH
IN CORKSCREW SWAMP**

Mizell prairie	Water	1.61×10^6 bacteria per milliliter
(pH 4)	Sediment	8.10×10^9 bacteria per gram dry wt
Central marsh	Water	1.23×10^6 bacteria per milliliter
(pH 7)	Sediment	2.75×10^9 bacteria per gram dry wt

E. Incubation Procedures

First, 10 ml of water or water/sediment slurry (50:1, vol/vol) were incubated at 25°C with 10 mg of radiolabeled lignocellulose in sterile bottles equipped with gassing ports.[11] The bottles were aerated with sterile humidified air for 15 min every 48 to 72 h. Mineralization of the [^{14}C]lignocellulose was monitored by trapping the evolved $^{14}CO_2$ in a series of two scintillation vials containing liquid scintillation medium.[11] Traps were changed after each aeration and radioactivity was quantified by liquid scintillation spectrometry. Controls were killed with 5% formalin, and no $^{14}CO_2$ was evolved during the incubation period. All incubations were done with three to five replicates.

Okefenokee bacterial isolates were cultured at pH 4 in Standard Methods nutrient broth diluted 1:10 (vol/vol) with distilled water. Corkscrew isolates were similarly cultured at pH 7. After incubating the broth cultures in a shaker at 25°C for 48 h, the bacteria were centrifuged and then resuspended in fresh medium diluted 1:100 to prevent growth during rate studies. Rate studies were carried out over a maximum of 8 h and were accomplished by determining the rates of loss of 2,4-dichlorophenoxyacetic acid methyl ester (2,4-DME) from the dilute media over the pH range from 3.5 to 7.0. The addition of Okefenokee isolates to pH 7.0 media and Corkscrew isolates to pH 3.5 media was the first exposure of these cultures to these large pH variations. The effect of pH on 2,4-DME transformation rates was measured as pseudo-first-order rate coefficients according to published procedures on a similar compound.[15] Water samples were extracted with isooctane, and 2,4-DME was quantified by gas-liquid chromatography using an electron capture detector. The effect of pH on transformation rates of 2,4-DME by natural populations of Okefenokee microflora was tested by adjusting the pH of Okefenokee water samples with NaOH. Because of low bacterial concentrations, rate studies using natural water samples lasted several days. Standard plate counts and the log-linear loss of 2,4-DME indicated that no net growth occurred during the incubation period. The initial 2,4-DME concentration in all incubations was 100 μg 1^{-1}.

III. RESULTS

Bacterial abundance in water and sediment samples from the Okefenokee Swamp was similar to, and in the case of sediments somewhat higher than, bacterial abundance in samples from the Corkscrew Swamp (Table 1). These results corroborate earlier data indicating that microbial biomass in the acidic water and sediment of the Okefenokee Swamp is similar to values reported for microbial biomass in other circumneutral pH environments.[5,6] Fungal abundance was significantly higher in Okefenokee sediments relative to Corkscrew sediments, but the contribution of fungi to the degradation of lignocellulosic detritus in both Okefenokee and Corkscrew sediments was minimal relative to the contribution from bacteria.[16]

Data from a previous study of the effect of pH on rates of polysaccharide and lignin

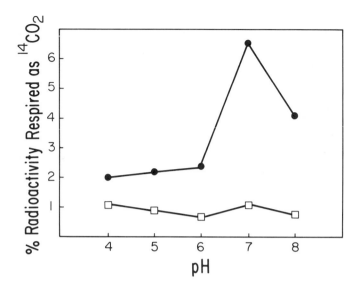

FIGURE 1.   Rates of mineralization of the polysaccharide (●) and lignin
(□) components of specifically [14C]-labeled lignocellulose in pH-adjusted
water from the Okefenokee Swamp (*in situ* pH 4).

mineralization in Okefenokee water are replotted in Figure 1.[4] The initial pHs are plotted
vs. the total mineralization of the polysaccharide and lignin components of specifically
radiolabeled lignocellulose during 10-d incubations. The pH of water samples was adjusted
up from pH 4 (*in situ* pH) with NaOH and allowed to stabilized for several days before
radiolabeled substrates were added. The pH remained fairly constant during the incubation
but did increase several tenths of a unit at pH 6 and 7. The pH optimum for polysaccharide
degradation was near pH 7, at which the rate of degradation was severalfold higher than at
pH 4. There was no clearly defined pH optimum for lignin degradation in Okefenokee water
(Figure 1); rates at pH 4 were the same as rates at pH 7.

In a similar experimental design, the pH optima for polysaccharide and lignin degradation
were determined in circumneutral pH water from Corkscrew Swamp. The pH of water
samples (pH 7) was adjusted with NaOH and HCl and allowed to stabilize for several days
before radiolabeled substrates were added. Most of the incubations maintained the adjusted
pH values during the 10-d incubation period, although incubations at pH 7 increased several
tenths of a unit and incubations at pH 9 decreased several tenths. As with Okefenokee
microflora, the pH optimum for polysaccharide degradation was near pH 7 (Figure 2). The
response of Corkscrew microflora to variations of pH was, however, quite different from
that observed with Okefenokee microflora. Rates of polysaccharide degradation decreased
gradually as the pH was decreased or increased from neutrality. In Okefenokee water the
rates of polysaccharide degradation decreased abruptly when the pH was decreased from
neutrality (Figure 1). The pH optimum for lignin degradation was also near 7 for Corkscrew
water samples (Figure 2).

These results suggest that natural microbial populations capable of rapid utilization of
lignocellulosic detritus will not develop in acidic environments. Moreover, it appears that
the microbial population in the naturally acidic water of the Okefenokee Swamp are no better
adapted for the degradation of polysaccharides than are microbial populations from circumneutral pH environments. An experiment was specifically designed to determine whether
Okefenokee microflora could mediate higher rates of polysaccharide and lignin degradation
in Okefenokee water than could Corkscrew microflora. Water samples from each environment were filter sterilized and divided into two equal portions (see Figure 3). Decaying

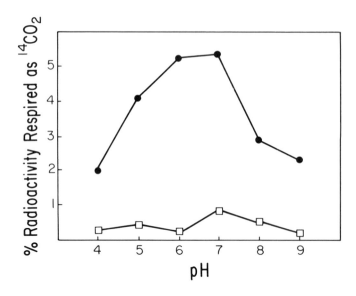

FIGURE 2.   Rates of mineralization of the polysaccharide (●) and lignin (□) components of specifically [¹⁴C]-labeled lignocellulose in pH-adjusted water from the Corkscrew Swamp (*in situ* pH 7).

MICROFLORA SWITCH

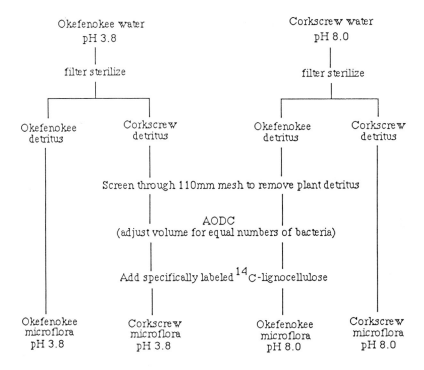

FIGURE 3.   Flow diagram of methods used to compare the lignocellulolytic potential of Okefenokee and Corkscrew microflora at different pHs.

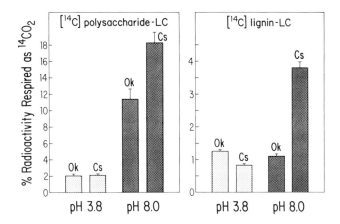

FIGURE 4. Comparison of the rates of mineralization of the polysaccharide and lignin components of specifically [^{14}C]-labeled lignocellulose by Okefenokee (Ok) and Corkscrew (Cs) microflora in Okefenokee (pH 3.8) and Corkscrew (pH 8.0) waters.

plant material and associated microbes from each environment were added as an inoculum to filter-sterilized water samples from both environments. The four water/detritus samples were gently shaken for 30 min, after which the plant material was removed and the bacterial abundance in all flasks adjusted so that all treatments had equivalent numbers of bacteria. Specifically [^{14}C]-labeled lignocellulose derived from the sedge, *C. walteriana*, was added, and samples were incubated for 10 d.

Rates of polysaccharide mineralization by Okefenokee and Corkscrew microflora were identical when incubated in pH 3.8 water from the Okefenokee Swamp (Figure 4). Rates of lignin mineralization by Okefenokee microflora were slightly higher than rates mediated by Corkscrew microflora when incubated in pH 3.8 water from the Okefenokee Swamp (Figure 4). Incubations with Okefenokee and Corkscrew microflora in pH 8.0 water from Corkscrew Swamp indicate that Corkscrew microflora were capable of mediating higher rates of polysaccharide and lignin mineralization than were Okefenokee microflora (Figure 4). These results agree with the pH optima data (Figures 1 and 2), which indicated a pH optima near 7 for polysaccharide degradation by both Okefenokee and Corkscrew microflora, and a pH optimum near 7 for lignin degradation by Corkscrew microflora but little difference between rates of lignin degradation at pH 4 and 8 by Okefenokee microflora.

In the above experiments, organic acids rather than mineral acids were responsible for variations in pH. We wanted to determine whether rates of lignocellulose degradation would be similarly affected by pH fluctuations due to mineral acids. Anthropogenic acidification can introduce various anions, such as sulfate and nitrate, that may also differentially affect microbial degradative processes. The effects of different mineral acids (HCl, H_2SO_4, HNO_3) and an organic acid (2,2-dimethyl succinic acid, DMSA) on rates of mineralization of uniformly [^{14}C]-labeled lignocellulose by natural populations of microflora were investigated at pHs 4.0, 5.5, and 7.0. As polysaccharides comprise approximately 95% of *Carex* lignocellulose and are degraded at higher rates than lignin, variations in rates of [U-^{14}C]lignocellulose degradation are due primarily to variations in rates of polysaccharide degradation. The tap water used in the 8-d incubations contained phosphate buffer but no added inorganic nitrogen source, and thus the microflora were nitrogen limited. The organic acid, DMSA, is relatively inert to microbial degradation and therefore did not serve as a major growth substrate in the incubations.

At pH 4, incubations with HCl or H_2SO_4 mediated the lowest rates of lignocellulose mineralization, and incubations with HNO_3 mediated four- to fivefold higher rates (Fig-

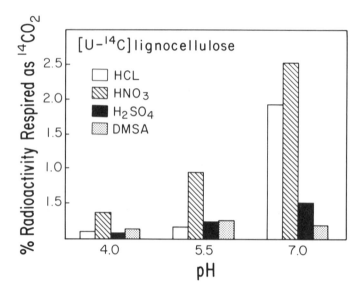

FIGURE 5.   Effects of pH and various types of anions on rates of
mineralization of uniformly [^{14}C]-labeled lignocellulose.

ure 5). Incubations with DMSA mediated lignocellulose mineralization rates that were approximately twofold higher than rates with HCl or H_2SO_4. At pH 5.5, incubations with HCl, H_2SO_4, or DMSA mediated similar rates of lignocellulose mineralization, whereas incubations with HNO_3 mediated four- to fivefold higher rates (Figure 5). In general, rates of lignocellulose mineralization at pH 5.5 were two- to threefold higher than rates at pH 4. At pH 7, rates of lignocellulose mineralization in incubations with HNO_3 were again the highest, and rates and were lowest in incubations with DMSA (Figure 5). Overall, rates of lignocellulose mineralization were 2- to 20-fold higher at pH 7 than at pH 4 for each of the acids. Notably, the organic acid had less overall effect on rates of lignocellulose degradation than the mineral acids. In this nitrogen-limited system the addition of HNO_3 had two major effects: decreasing the pH of incubations which decreased rates of lignocellulose degradation, and increasing the available nitrogen to microflora thereby stimulating rates of lignocellulose degradation. The change in pH from 7 to 4 caused a sevenfold decrease in lignocellulose mineralization rates in incubations with HNO_3. The nitrogen stimulation of microbial degradation increased the rates of lignocellulose degradation by approximately fivefold above rates in incubations with H_2SO_4 at the various pHs.

Previous studies of the rates of microbial utilization of dissolved substrates by natural populations of Okefenokee microflora indicated that, unlike the data presented for lignocellulose degradation, rates were not depressed in the low pH water of the Okefenokee Swamps.[5] These data suggest that rates of microbial utilization of dissolved substrates are not dependent on pH over the range from 4 to 7. To further investigate the pH dependence of microorganisms for the utilization of dissolved substrates, several pure cultures of bacteria were isolated from Okefenokee and Corkscrew water and the relationship between pH and rates of degradation of the xenobiotic, 2,4-DME, was determined. Of the four bacterial isolates from the Okefenokee Swamp, only *Pseudomonas cepacia* (#1465) demonstrated a slight increase in rates of degradation of the xenobiotic with increasing pH over the range from 3.5 to 7.0 (Figure 6). Degradation rates of 2,4-DME by natural populations of Okefenokee microflora in pH-adjusted Okefenokee water were not affected by pH over the range from 3.5 to 7.0 (data not shown). Similarly, bacterial strains from Corkscrew Swamp exhibited little, if any, variation in rates of degradation over the same pH range (Figure 7). These results are consistent with previous studies[5] and suggest that differences in the mech-

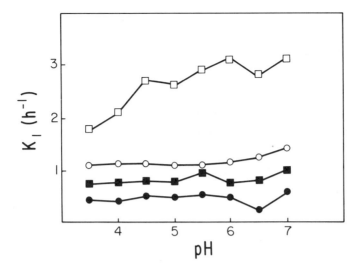

FIGURE 6. Effects of pH on first-order decay constants (K_1) for the degradation of 2,4-DME by bacteria isolated from the Okefenokee Swamp. (□) *Pseudomonas cepacia* #1465, (○) *P. cepacia* #1473, (■) *P. cepacia* #1467, and (●) *Chromobacterium violacium* #1466.

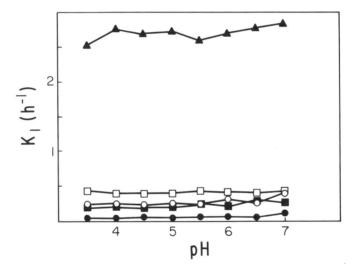

FIGURE 7. Effects of pH on first-order decay constants (K_1) for the degradation of 2,4-DME by bacteria isolated from the Corkscrew Swamp. (▲) *Pseudomonas maltophila* #1486, (□) *P. cepacia* #1468, (○) *P. vesicularis* #1472, (■) *Plesiomonas shigelloides* #1471, and (●) *Pseudomonas cepacia* #1469.

anisms (i.e., intracellular vs. extracellular) of microbial degradation of dissolved and particulate organic substrates are responsible for the different responses to pH variations.

IV. DISCUSSION

In many freshwater ecosystems, vascular plant detritus derived from allochthonous inputs of riparian vegetation or autochthonous production of aquatic vegetation provides a major source of nutrients and energy to aquatic food webs.[17] This material is comprised predominantly of structural polysaccharides and lignin and is commonly referred to as lignocellulosic detritus.[12] A variety of field and laboratory studies have indicated that rates of decomposition of lignocellulosic detritus are lower at low pHs relative to circumneutral pHs.[4,18-24] Several of these studies have suggested that the observed reduction in rates of lignocellulosic detritus at low pHs is an indirect effect of acidic conditions, such as a reduction in invertebrate populations.[20,24,25] Undoubtedly these indirect effects of acidification are important to decomposition processes; however, the studies presented herein demonstrate that the enzymatically mediated degradation of lignocellulosic detritus by microorganisms is itself directly inhibited at low pHs.

A sharp distinction also appears to exist between the effects of acidity on the microbial utilization of insoluble polymers, such as polysaccharides, and the microbial utilization of dissolved substrates. Acid stress has a much more marked effect on the microbial degradation of polymers; water-soluble compounds appear to be degraded by bacteria at similar rates over the pH range from 3.5 to 7.0. Large polymeric substrates cannot be directly taken up by bacteria and fungi, but must first be degraded by exoenzymes to low molecular weight dissolved compounds that are then transported into the cell. Exoenzymes required for polymer cleavage are thereby exposed to the aqueous environment outside the cell, whereas compounds that can be transported inside the cell are degraded within the highly regulated cytoplasm. We suggest that the reduced rates of degradation of lignocellulosic detritus in low pH waters are due, in large part, to some inhibition of exoenzymes, perhaps a conformational change, that slows polymer cleavage. Furthermore, we suggest that rates of microbial degradation of other polymeric substrates, such as chitin, will also be reduced at low pHs.

Several investigators have noted that fungal biomass is often greater in acidic than in circumneutral pH environments, suggesting that acidic conditions favor fungal growth and activity.[16,22,24-26] At the same time, these and other studies have found that rates of microbial degradation of lignocellulosic substrates were much lower under acidic conditions. Fungi are considered to be the major decomposers of lignocellulosic substrates in terrestrial systems and have been shown to produce cellulolytic enzymes with acidic pH optima.[27] It is, therefore, somewhat surprising to find that rates of degradation of lignocellulosic detritus decrease upon acidification even though fungal biomass increases. Treatment of acidified waters with lime increases the rates of degradation of lignocellulosic detritus and greatly decreases fungal biomass, further suggesting a minor role of fungi in lignocellulosic detritus and greatly decreases fungal biomass, further suggesting a minor role of fungi in lignocellulose degradation in aquatic environments.[26] Direct investigations of the relative contributions of bacteria and fungi to the degradation of lignocellulosic detritus in a variety of aquatic environments including the acidic waters of the Okefenokee Swamp indicate that bacteria, rather than fungi, are the predominant degraders of polysaccacharides and lignin.[16] It is evident that fungi are abundant in acidified waters, but to our knowledge, it has not been shown that fungal activity is actually higher under acidic conditions. The greater biomass of fungi in low pH systems may result from a decrease in populations of invertebrate predators. Further research is needed to elucidate the role of fungi in decomposition processes in acidified aquatic environments.

Short-term laboratory studies have yielded valuable information on the responses of microorganisms to acid stress, but have been criticized for not reflecting the resilience of natural populations.[28] Laboratory studies of microbial processes in water and sediment samples from naturally acidic environments alleviate many of the shortcomings of laboratory experiments and can be helpful in determining mechanisms of inhibition in acid-stressed environments. Application of results from such laboratory experiments can help explain results from field studies where mechanisms are more difficult to ascertain, whereas field studies are needed to verify laboratory results. As an example for the need of field verification of laboratory results, we refer to the long-term studies of acid stress on lake ecosystems carried out by Schindler et al.[28] During long-term whole-lake acidification experiments (Lake 223), rates of decomposition in sediments were unaffected over the pH range of 6.7 to 5.1 even though laboratory experiments in which sediment pore-water pH was strictly maintained indicated that decomposition was inhibited at lower pHs.[29] Measurements of sediment pH indicated that, unlike in the aerobic water column, the pH of lake sediments had not decreased. Apparently, the activity of sulfate-reducing bacteria in the sediments maintained a circum-neutral pH within millimeters of the sediment surface.[29] Without field measurements of pH in sediments, the acid-neutralizing capability of anaerobic bacteria could have been over-looked.

Inorganic nitrogen concentrations are limiting to primary and secondary production in many aquatic ecosystems,[30] and thus atmospheric precipitation containing nitric acid may actually stimulate certain processes.[31] Bacterial growth efficiencies on lignocellulosic detritus are often limited by available nitrogen as these substrates typically have carbon/nitrogen ratios that are greater than 40, whereas bacteria have carbon/nitrogen ratios near 5.[32] Additions of inorganic nitrogen to natural water samples often increase overall rates of degradation of lignocellulosic detritus and bacterial growth efficiencies.[33-35] We observed a similar effect of nitric acid on rates of lignocellulose degradation in comparison to the effects of other mineral and organic acids. The inhibitory pH effect of nitric acid on lignocellulose degradation rates was greatly dampened by the stimulatory effect of the added nitrogen. The effects of acid stress on decomposition processes will be very dependent, therefore, on the types and concentrations of anions as well as the concentration of the hydrogen ion deposited during atmospheric precipitation.

Aquatic ecosystems in which microbial decomposition and nutrient regeneration may be most affected by anthropogenic acidification are those that have well-mixed waters and an abundance of particulate organic detritus. Thus, microbial decomposition processes in streams that receive large inputs of riparian vegetation, in wetlands, and in shallow lakes with a high biomass of rooted vegetation may be particularly susceptible to acid stress. Microorganisms and the processes they mediate are highly resilient, but it appears that microbial decomposition processes that rely on exoenzymes are particularly susceptible to inhition by acid stress. Moreover, it appears that microbial populations capable of mediating rates of polysaccharide degradation that are comparable to rates in circumneutral environments will not develop, as such populations were not detected in environments that have been acidic for thousands of years.

ACKNOWLEDGMENTS

We thank the staff of the Okefenokee National Wildlife Refuge for their cooperation during this study. We also thank M. J. Duever and the National Audubon Society for permission to collect samples from Corkscrew Swamp Sanctuary and W. J. Wiebe and W. C. Steen for comments on an earlier version of this manuscript. This work was supported in part by grants BSR 8215587 and BSR 8114823 from the National Science Foundation.

REFERENCES

1. **Fenchel, T. and Blackburn, T. H.,** *Bacteria and Mineral Cycling,* Academic Press, London, 1979.
2. **Azam, F., Fenchel, T., Field, J. G., Gray, J. S., Meyer-Reil, L. A., and Thingstad, F.,** The ecological role of water-column microbes in the sea, *Mar. Ecol. Prog. Ser.,* 10, 257, 1983.

3. **Hunt, H. W., Coleman, D. C., Cole, C. V., Ingham, R. E., Elliott, E. T., and Woods, L. E.,** Simulation model of a food web with bacteria, amoebae, and meatodes in soil, in *Current Perspectives in Microbial Ecology,* Klug, M. J. and Reddy, C. A., Eds., American Society for Microbiology, Washington, D.C., 1984, 346.

4. **Benner, R., Moran, M. A., and Hodson, R. E.,** Effects of pH and plant source on lignocellulose biodegradation rates in two wetland ecosystems, the Okefenokee Swamp and a Georgia salt marsh, *Limnol. Oceanogr.,* 30, 489, 1985.

5. **Murray, R. E. and Hodson, R. E.,** Microbial biomass and utilization of dissolved organic matter in the Okefenokee Swamp ecosystem, *Appl. Environ. Microbiol.,* 47, 685, 1984.

6. **Moran, M. A., Maccubbin, A. E., Benner, R., and Hodson, R. E.,** Dynamics of microbial biomass and activity in five habitats of the Okefenokee Swamp ecosystem, *Microb. Ecol.,* in press.

7. **Havas, M., Hutchinson, T. C., and Likens, G. E.,** Red herrings in acid rain research, *Environ. Sci. Technol.,* 18, 176, 1984.

8. **Reuss, J. O. and Johnson, D. W.,** *Acid Deposition and the Acidification of Soils and Waters,* Springer-Verlag, New York, 1986.

9. **Feeny, P. P.,** Inhibitory effect of oak leaf tannins on the hydrolysis of proteins by trypsin, *Phytochemistry,* 8, 2119, 1969.

10. **Benoit, R. E. and Starkey, R. L.,** Enzyme inactivation as a factor in the inhibition of decomposition of organic matter by tannins, *Soil Sci.,* 105, 203, 1968.

11. **Benner, R., Maccubbin, A. E., and Hodson, R. E.,** Preparation, characterization, and microbial degradation of specifically radiolabeled [^{14}C]lignocelluloses from marine and freshwater macrophytes, *Appl. Environ. Microbiol.,* 47, 381, 1984.

12. **Benner, R., Newell, S. Y., Maccubbin, A. E., and Hodson, R. E.,** Relative contributions of bacteria and fungi to rates of degradation of lignocellulosic detritus in salt-marsh sediments, *Appl. Environ. Microbiol.,* 48, 36, 1984.

13. **Hobbie, J. E., Daley, R. J., and Jasper, S.,** Use of Nuclepore filters for counting bacteria by fluorescence microscopy, *Appl. Environ. Microbiol.,* 33, 1225, 1977.

14. **Hodson, R. E., Moran, M. A., Lewis, D. L., Murray, R., Teska, J. D., and Benner, R.,** Microbial biogeochemical processes in a naturally acidic wetland, the Okefenokee Swamp, in *Microbial Interactions in Acid-Stressed Aquatic Ecosystems,* Rao, S. S., Ed., CRC Press, Boca Raton, FL, 1989, chap. 4.

15. **Lewis, D. L., Kollig, H. P., and Hall, T. L.,** Predicting 2,4-dichlorophenoxyacetic ester transformation rates in periphyton-dominated ecosystems, *Appl. Environ. Microbiol.,* 46, 146, 1983.

16. **Benner, R., Moran, M. A., and Hodson, R. E.,** Biogeochemical cycling of lignocellulosic carbon in marine and freshwater ecosystems: relative contributions of procaryotes and eucaryotes, *Limnol. Oceanogr.,* 31, 89, 1986.

17. **Webster, J. R. and Benfield, E. F.,** Vascular plant breakdown in freshwater ecosystems, *Annu. Rev. Ecol. Syst.,* 17, 567, 1986.

18. **Brock, T. C. M., Boon, J. J., and Paffen, B. G. P.,** The effects of the season and of water chemistry on the decomposition of *Nymphaea alba* L; weight loss and pyrolysis mass spectrometry of the particulate matter, *Aquat. Bot.,* 22, 197, 1985.

19. **Burton, T. M., Stanford, R. M., and Allan, J. W.,** Acidification effects on stream biota and organic matter processing, *Can. J. Fish. Aquat. Sci.,* 42, 669, 1985.

20. **Carpenter, J., Odum, W. E., and Mills, A.,** Leaf litter decomposition in a reservoir affected by acid mine drainage, *Oikos,* 41, 165, 1983.

21. **Hildrew, A. G., Townsend, C. R., Francis, J., and Finch, K.,** Cellulolytic decomposition in streams of contrasting pH and its relationship with invertebrate community structure, *Freshwater Biol.,* 14, 323, 1984.

22. **Hoeniger, J. F. M.,** Decomposition studies in two central Ontario lakes having surficial pHs of 4.6 and 6.6, *Appl. Environ. Microbiol.,* 52, 489, 1986.

23. **McKinley, V. L. and Vestal, J. R.,** Effects of acid on plant litter decomposition in an Arctic Lake, *Appl. Environ. Microbiol.,* 43, 1188, 1982.

24. **Traaen, T. S.,** Effects of acidity on decomposition of organic matter in aquatic environments, in Proc. Int. Conf. Ecological Impact of Acid Precipitation, Drablos, D. and Tollan, A., Eds., SNSF Project, Oslo-Aas, Norway, 1980, 340.

25. **Hall, R. J., Likens, G. E., Fiance, S. B., and Hendrey, G. R.,** Experimental acidification of a stream in the Hubbard Brook experimental forest, New Hampshire, *Ecology,* 61, 976, 1980.

26. **Hendrey, G., Baalsrud, K., Traaen, T. S., Laake, M., and Raddum, G.,** Acid precipitation: some hydrobiological changes, *Ambio,* No. 5-6, 224, 1976.

27. **Ljungdahl, L. G. and Eriksson, K. E.,** Ecology of microbial cellulose degradation, *Adv. Microb. Ecol.,* 8, 237, 1985.

28. **Schindler, D. W., Mills, K. H., Malley, D. F., Findlay, D. L., Shearer, J. A., Davies, I. J., Turner, M. A., Linsey, G. A., and Cruikshank, D. R.,** Long-term ecosystem stress: the effects of years of experimental acidification on a small lake, *Science,* 228, 1395, 1985.
29. **Kelly, C. A., Rudd, J. W. M., Furutani, A., and Schindler, D. W.,** Effects of lake acidification on rates of organic matter decomposition in sediments, *Limnol. Oceanogr.,* 29, 687, 1984.
30. **Mann, K. H.,** *Ecology of Coastal Waters,* University of California Press, Berkeley, 1982.
31. **Paerl, H. W.,** Enhancement of marine primary production by nitrogen-enriched acid rain, *Nature,* 316, 747, 1985.
32. **Linley, E. A. S. and Newell, R. C.,** Estimates of bacterial growth yields based on plant detritus, *Bull. Mar. Sci.,* 35, 409, 1984.
33. **Aumen, N. G., Bottomley, P. J., Ward, G. M., and Gregory, S.,** Impact of nitrogen and phosphorus on [^{14}C]lignocellulose decomposition by stream wood microflora, *Appl. Environ. Microbiol.,* 49, 1113, 1985.
34. **Benner, R., Lay, J., K'nees, E., and Hodson, R. E.,** Efficiency of bacterial growth on lignocellulose: implications for detritus-based food webs, *Limnol. Oceanogr.,* submitted.
35. **Newell, R. C., Linley, E. A. S., and Lucas, M. I.,** Bacterial production and carbon conversion based on saltmarsh plant debris, *Estuarine Coastal Shelf Sci.,* 17, 405, 1983.

Chapter 4

MICROBIAL BIOGEOCHEMICAL PROCESSES IN A NATURALLY ACIDIC WETLAND, THE OKEFENOKEE SWAMP

Robert E. Hodson, Mary Ann Moran, David L. Lewis, R. Murray, Jeffrey D. Teska, and Ronald Benner

TABLE OF CONTENTS

I. PHYSICAL AND CHEMICAL CHARACTERISTICS

The Okefenokee Swamp, in southeastern Georgia and northern Florida, is one of the largest freshwater wetlands in the U.S. (Figure 1). It is an acidic, peat-accumulating environment consisting of a mosaic of aquatic habitats including forested and shrub swamps, and water lily and emergent macrophyte prairies, interspersed with dry pine islands. Thus, the Okefenokee presents a unique natural laboratory for studying the long-term biogeochemical and trophodynamic effects of, and biological adaptation to, system acidification. If microbial populations anywhere (or the biogeochemical processes they mediate) can become adapted gradually to acidic conditions, such populations should be found in the Okefenokee Swamp which is assumed to have become acidified slowly by natural accumulation of organic acids and to have been acidic for the past 8000 to 10,000 years. If, on the other hand, acid-tolerant microbial populations are not found in environments such as Okefenokee, it seems doubtful that they will be widespread features of other naturally or anthropogenically acidified aquatic ecosystems.

II. IMPORTANCE OF MICROBIAL PROCESSES TO SYSTEM FUNCTION AND STRUCTURE

The overall natural history of this unique ecosystem has been reviewed thoroughly in a volume edited by Cohen et al.[1] In this system, primary production is dominated by vascular plants with some, as yet unquantified, contributions from mosses, and epiphytic and planktonic algae.[1] Apparently, very little of the vascular plant biomass in this and most other wetlands is grazed directly by animals,[2] a fact which has been attributed to its being composed primarily of refractory, low-nutrient, lignocellulose, the macromolecular complex of lignin and polysaccharides that comprises the structural components of higher plants.

Rather than being grazed, most of the plant biomass dies and becomes a component of the abundant organic detritus and peat of the swamp sediments and water column. Consequently, previous investigators have assumed that microbial transformations of the major reservoirs of organic matter (dissolved organic compounds as well as particulate detritus and peat) are important mechanims controlling both nutrient availability and the conversion of relatively refractory organic matter into highly assimilable (to animals) microbial biomass. Thus, the potential effects of the highly acid conditions on microbial processes would be expected to have a major impact on the animal food web as well. However, until recently, little information has been available regarding either the rates of microbially mediated processes in the Okefenokee or the factors that control microbial secondary production or transformation and dissimilation of organic matter. Likewise, although confirming data were not available, the assumption was made that biomass of detritivores and other metazoans in such systems were composed of carbon derived ultimately from the biomass of vascular plants after its transformation by the aquatic microflora.[2]

III. MICROBIAL ECOLOGY OF THE OKEFENOKEE WETLAND ENVIRONMENTS

During the past several years, we have been conducting studies of the microbial processes important to the Okefenokee Swamp ecosystem as part of the National Science Foundation-sponsored Long Term Ecological Research (LTER) program. Our goal has been both to characterize the system microbiologically and to gather data on the overall role of microbial biomass and microbial processing of particulate and dissolved organic matter for use by us and other LTER scientists in overall models of swamp structure and function. In the process we have uncovered some interesting and, perhaps, surprising facets of microbial activity in

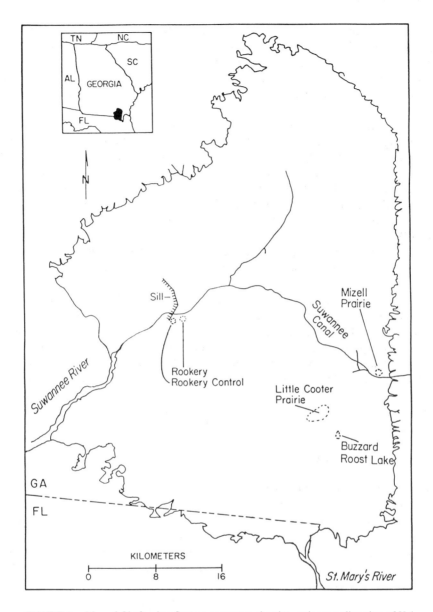

FIGURE 1.   Map of Okefenokee Swamp ecosystem showing major sampling sites of University of Georgia research team.

this highly acidic, "detritus-based" ecosystem. Our studies revealed that the microbial processes operative in the Okefenokee differ both qualitatively and quantitatively from those in other wetland ecosystems in a number of ways. We provide here a brief review of the current status of our knowledge of microbial biomass and activity in the Okefenokee as well as the initial presentation of some recently acquired new data. We believe that the baseline studies done to date are now sufficient to supply the needed background for continued experimental studies designed specifically to compare this naturally acidic ecosystem with others undergoing acidification due to acid rain or other anthropogenic factors.

IV. INITIAL STUDIES

Much of the work conducted during this time required the modification of existing methodologies or development of completely new procedures for measuring microbial biomass and rates of degradative processes in the highly acidic (pH = 3.5) waters and peat of the Okefenokee.[3-6] Considerable time was spent also in development, optimization, and utilization of new methods for specifically radiolabeling the lignin and polysaccharide components of swamp macrophytes for use in experiments to evaluate the factors that control rates of microbial transformation and dissimilation of detrital lignocelluloses and the resulting bacterial secondary production of biomass.

Our initial work in the Okefenokee was guided in part by results of previous studies of the microbiology of acidic peat bogs, which suggested that these systems exhibit reduced microbial biomass and activity compared with circumneutral water, soils, and sediments.[7] The combination of low pH, low redox potential, and the refractory nature of the available organic substrates were thought to limit both microbial numbers and activity in these environments.[8-9] On the basis of previous studies then, one might expect that microbial biomass and the rates of microbially mediated processes, such as production of particulate organic material (POM) and turnover of dissolved and particulate organic compounds in sediment and water, would be lower in the Okefenokee Swamp than in circumneutral pH aquatic ecosystems. However, our first investigations of Okefenokee microbial populations indicated that this may not be the case.

We undertook extensive surveys of microbial biomass and activity in five representative habitat types in the swamp including sedge prairie (Mizell Prairie), water lily prairie (Little Cooter Prairie), open lake water (Buzzards Roost Lake), and a large bird rookery and adjacent control area in a cypress swamp (Rookery and Rookery Control) (shown in Figure 1). These studies revealed significant spatial and seasonal variability in microbial biomass and activity both within and among habitats. However, in general our results revealed that, unlike the better-studied northern peat bogs, the Okefenokee is not depauperate with respect to microbial populations[7] but instead supports biomass and rates of some microbially mediated processes that equal or exceed those in neutral-pH wetlands.[3-5]

V. PROCESSES INDEPENDENT OF pH

A. Biomass and Secondary Production

Values for microbial biomass measured as particulate adenosine triphosphate in water and sediments were found to be high relative to values for other previously studied wetland and soil environments as can be seen from Table 1 taken from a paper published in 1984 by Murray and Hodson,[3] as were values for bacterial biomass measured by direct microscopic counts using epifluorescence-stained samples.[10] Bacterial biomass, however, varies by 15-fold or more over the average year. Overall, bacterial secondary production, estimated as incorporation of thymidine into bacterial DNA, was also high relative to other organically rich aquatic ecosystems that have been examined but varied seasonally more than 100-fold, as is evident from Figure 2 reprinted from a 1985 paper by Murray and Hodson.[4] Peak values for secondary production and bacterial numbers were found to coincide with periods of maximal production of vascular plants.[4,5,10] Clearly, the Okefenokee Swamp supports levels of microbial biomass and bacterial secondary production in the water column and surface sediments similar to those in other organically rich aquatic ecosystems.[3,10]

B. Bacterial Uptake of Simple Organic Compounds

Radiotracer uptake experiments indicated that bacterioplankton populations support relatively high rates of turnover of labile dissolved organic substrates such as D-glucose[3] (Table 2, taken from Murray and Hodson[3]) although these data, per se, do not exclude the possibility that, although rates are high, pH might still be limiting the rates of turnover of dissolved compounds. To examine this possibility, we recently conducted experiments to determine what direct effect, if any, pH has on rates of turnover of dissolved organic compounds in

Table 1

**RANGES OF ATP CONCENTRATIONS IN
THE WATER COLUMN AND SURFACE
SEDIMENTS OF THE OKEFENOKEE
SWAMP AND OTHER WETLANDS AND IN
TERRESTRIAL SOILS**

Location	ATP conc	Ref.
Water column		
South Carolina salt marsh	0.3—3.5	20
Cheseapeake Bay salt marsh	0.02—2.7	21
Sapelo Island salt marsh, GA	1.0—4.0	22
Okefenokee Swamp, GA	0.3—6.6	23
Sediments		
Long Island Sound, NY	1.1—7.6	24
Sapelo Island salt marsh, GA	1.2—9.8	25
13 Australian soil types	0.6—9.0	26
9 New Zealand grassland soils	2.2—10.7	27
Freshwater marsh, MI	5.3—16.7	28
Okefenokee Swamp, GA	1.0—28.0	23

Note: The water column concentrations are given in micrograms per liter. Sediment concentrations are given in
micrograms per gram (dry weight).

From Murray, R. E. and Hodson, R. E., *Appl. Environ. Microbiol.*, 47, 685, 1984. With permission.

the naturally acidic Okefenokee as well as in Big Cypress Swamp (south Florida) which has
pH near neutrality. This was addressed by incubating ^{14}C-labeled D-glucose or commercially
available mixtures of ^{14}C-radiolabeled amino acids added at trace levels with samples of
unamended Okefenokee or Big Cypress Swamp water containing the natural microbial
assemblages, or to Okefenokee or Big Cypress Swamp water that had been amended by
artificially increasing the pH at intervals up to 7. Samples were preincubated at the amended
pH for periods of up to 11 d prior to addition of the tracer; then the natural populations
were allowed to take up the substrate for short intervals (2 to 5 h following procedures
worked out previously.[3])

Interestingly, little dependence of dissolved substrate uptake on sample pH was observed
(Figures 3A to D). D-Glucose uptake into microbial biomass was not related significantly
to pH in either the Okefenokee or the Big Cypress Swamp water samples. Likewise, uptake
of the radiolabeled amino acid mixture was independent of pH over the range of 3.8 to 7.0
in Big Cypress water. However, in Okefenokee Swamp water samples, the amino acid
mixture uptake was consistently higher (up to 2.5-fold higher) at pH 7 than at pH 4, 5 or
6 but only after preincubations at elevated pH for 7 d or longer.

VI. PROCESSES THAT ARE pH DEPENDENT

A. pH Dependence of Okefenokee Bacterial Isolate Growth Rate

The fact that utilization of readily degradable dissolved organic substrates, such as
D-glucose and, for the most part, L-amino acids, by the natural bacterial assemblages in the
Okefenokee (as well as in the circumneutral waters of Big Cypress Swamp) was found to
be independent of pH over the environmentally significant pH range could be interpreted as
indicating the presence of acidophilic or acid-tolerant bacteria in these wetlands. However,
attempts to isolate such organisms have, thus far, been unsuccessful. In an attempt to obtain

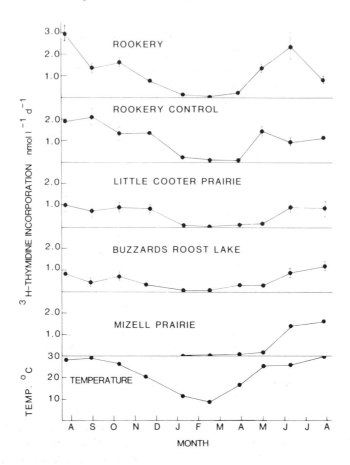

FIGURE 2.   Annual cycle of thymidine incorporation into cold-TCA-insoluble material in five aquatic habitats of the Okefenokee Swamp, August 1982 to August 1983. (From Murray, R. E. and Hodson, R. E., *Appl. Environ. Microbiol.*, 49, 650, 1985. With permission.)

representatives of such organisms, bacteria were isolated on plates prepared of nutrient agar or 0.1% nutrient agar adjusted to either pH 4 (Okefenokee water samples) or pH 7 (Big Cypress Swamp water samples). Colonies that grew after 24-h incubation at 30°C were picked randomly and maintained on identical medium from which they were isolated at the respective pH of isolation. All isolates were screened on Triple Sugar Iron (TSI) and all nonfermenters were identified using API NFT strips (API Analytab Products, Plainview, NY). All isolates were also screened with bromcresol green O-F glucose at pH 4. All other isolates were identified using conventional methods (Table 2).

A total of 21 distinct isolates (14 from Okefenokee and 7 from Big Cypress Swamp) were examined for growth rates supported on brain heart infusion (BHI) broth at pH 7 and pH 4. Cultures were pregrown in broth tubes for 48 h at 30°C, centrifuged, and washed twice with sterile 0.85% saline, then resuspended and adjusted to a final optical density (OD) of 0.500 at 660 nm. Of each suspension, 50 μl was then added to incubation cartridges (MS-2) containing BHI broth at pH 4 or pH 7 and growth rates were computed from optical density measurements made automatically at 5-min intervals over 36 h.

Of the 21 strains, 19 grew at significant rates at pH 7 as indicated by the values for slopes of OD vs. time, although some underwent lag periods before detectable growth occurred. None of the isolates supported detectable growth as measured by OD increase during the 36-h incubation when incubated at pH 4, notwithstanding the fact that the majority had been

Table 2

TURNOVER TIMES[a] FOR DISSOLVED D-GLUCOSE IN THE WATER COLUMN AND SURFACE SEDIMENTS OF FOUR OKEFENOKEE SWAMP HABITATS, 1980

Habitat	September	December	February	March	April	May	June	July	August	Mean
Water column										
Chesser Prairie	244 ± 38	66 ± 6	5 ± 0.3	2 ± 0.2	8 ± 4	16 ± 2	—	—	1 ± 0.04	49 ± 89
Chesser Shrub	40 ± 1	701 ± 252	38 ± 4	15 ± 1	9 ± 1	—	—	—	—	161 ± 303
Grand Cypress	—	510 ± 31	106 ± 7	9 ± 1	—	—	—	—	—	42 ± 56
Mizell Prairie	260 ± 13	54 ± 43	12 ± 5	3 ± 1	106 ± 13	—	—	—	—	87 ± 105
Surface sediment										
Chesser Pr.	2 ± 0.4	5 ± 1	5 ± 1	13 ± 6	26 ± 7	9 ± 3	2 ± 1	3 ± 0.4	6 ± 1	8 ± 8
Chesser Shr.	3 ± 1	24 ± 3	11 ± 2	8 ± 2	8 ± 3	8 ± 4	8 ± 2	6 ± 1	2 ± 0.06	9 ± 6
Grand Cypress	8 ± 1	70 ± 27	7 ± 0.4	13 ± 5	10 ± 2	9 ± 2	*	5 ± 2	2 ± 0.4	16 ± 22
Mizell Prairie	3 ± 0.2	5 ± 1	6 ± 1	23 ± 2	24 ± 2	8 ± 3	5 ± 1	5 ± 0.1	2 ± 0.4	9 ± 8

[a] The water column turnover times are given in hours; surface sediment turnover times are given in minutes. —, No standing water and *, not sampled.

From Murray, R. E. and Hodson, R. E., *Appl. Environ. Microbiol.*, 47, 685, 1984. With permission.

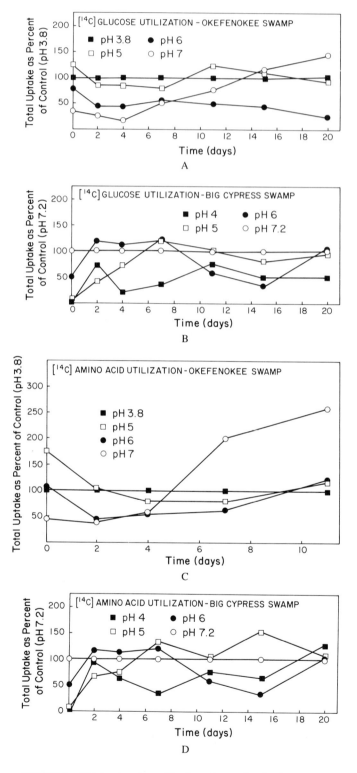

FIGURE 3.   Uptake at various pH values of tracer levels of radiolabeled D-glucose or L-amino acid mixtures from water of (A and C) Okefenokee Swamp or (B and D) Big Cypress Swamp by the natural microbial populations of each environment. Sample pH was adjusted by addition of acid or base to natural water samples.

Table 3
IDENTITIES, GROWTH RATES (EXPRESSED AS SLOPES OF GROWTH CURVES), AND LAG TIMES EXHIBITED BEFORE GROWTH ON BRAIN HEART INFUSION BROTH OF BACTERIAL STRAINS ISOLATED FROM OKEFENOKEE SWAMP OR CORKSCREW SWAMP (BIG CYPRESS SWAMP) AND INCUBATED AT pH 7

Isolate number	Species	Slope	Lag time (h)	Sources
1472	*Pseudomonas vesicularis*	2.56	0.25	Corkscrew water at pH 7
1473	*P. cepacia*	3.57	17.0	Okefenokee aufwuchs at pH 4
1474	— [a]	2.71	12.0	Okefenokee surface — shallow water at pH 4
1477	*Pseudomonas fluorescens*	4.54	1.0	Corkscrew water at pH 7
1478	*P. cepacia*	2.56	13.0	Okefenokee aufwuchs at pH 4
1479	*P. cepacia*	3.64	10.0	Okefenokee surface — shallow water at pH 4
1481	*Micrococcus* sp.	4.50	2.0	Corkscrew water at pH 7
1483	*Pseudomonas cepacia*	3.57	3.0	Okefenokee surface — shallow water at pH 4
1484	— [b]	1.06	12.0	Okefenokee deep water at pH 4
1486	*Pseudomonas maltophilia*	0	0	Corkscrew water at pH 7
1487	*P. cepacia*	2.59	7.0	Okefenokee aufwuchs at pH 4
1488	*P. cepacia*	2.63	11.0	Okefenokee surface — shallow water at pH 4
1490	*P. paucimobilis*	0	0	Okefenokee sediment
1491	NFT #0071040	0.74	0.25	Corkscrew water at pH 7
1492	— [c]	0.81	23.0	Okefenokee surface — shallow water at pH 4
1493	CDC grp V E-1	2.83	9.0	Okefenokee deep water at pH 4
1495	*Pseudomonas fluorescens*	4.73	2.0	Corkscrew water at pH 7
1496	— [d]	0.85	16.0	Okefenokee aufwuchs at pH 4
1497	*Pseudomonas cepacia*	3.07	19.0	Okefenokee surface — shallow water at pH 4
1498	—	2.88	13.0	Corkscrew water at pH 7
1499	— [e]	1.28	17.0	Okefenokee surface — shallow water at pH 4

Note: No growth was detected when strains were grown at pH 4 (i.e., slopes were equal to zero).

[a] Bromcresol green O-F glucose = oxidizer, TSI = a/N, cat+, gm+, rod.
[b] Bromcresol green O-F glucose = oxidizer, oxidase.
[c] Bromcresol green O-F glucose = oxidizer, oxidase.
[d] Bromcresol green O-F glucose = oxidizer, oxidase.
[e] Bromcresol green O-F glucose = oxidizer, oxidase.

isolated from Okefenokee water or sediment at approximately pH 4 (Table 3). Thus, at least for easily isolated strains of *Pseudomonas,* we have not observed an obvious adaptation toward acid tolerance in the Okefenokee ecosystem; rather, the Okefenokee heterotrophic bacterial strains isolated to date do not differ from those of neutral-pH environments. However, pure-culture experiments conducted thus far were conducted using organically rich nutrient broth or agar and have not simulated Okefenokee *in situ* conditions of organic nutrient quality or concentration or source of acidity (Okefenokee pH is set at approximately 4 by the presence of organic acids rather than mineral acids). Thus, further experimentation will be necessary to reconcile the observations that Okefenokee bacterial isoaltes do not appear to grow at pH 4 while certain bacterially mediated processes in the Okefenokee proceed at high rates relative to other wetland ecosystems.

B. Transformation of Lignocellulosic Detrital Material in the Okefenokee

In contrast to the high rates of bacterial production and uptake and dissimilation of labile dissolved substrates in the Okefenokee, microbial degradation of particulate lignocellulosic material is, in fact, depressed in the Okefenokee relative to other wetland ecosystems. During the initial 2 years of our study, we modified and optimized procedures for preparing and characterizing (chemically and radiochemically) specifically radiolabeled [14]C-(lignin) and [14]C-(polysaccharide)-lignocelluloses[11] from a wide range of freshwater, marine, and aquatic vascular plants and details of the procedures developed can be found in our previous papers.[6,12] Labeled lignocellulose, prepared from live plant cuttings incubated in light with labeled lignin or polysaccharide precursor compounds, is composed of lignin and polysaccharide having a macromolecular matrix (and hence degradability) that is characteristic of the particular plant species and physiological state. After extraction and characterization, these labeled lignocelluloses can be used in laboratory or field incubations effectively to determine rates of mineralization, solubilization, or conversion to microbial biomass of lignocellulosic detrital material; with carefully designed experiments, such incubations can be used to examine independently and with high analytical sensitivity the effects of various environmental controlling factors (such as pH) on the specific rates of transformation of individual molecular components (cellulose, hemicellulose, and lignin) of detritus and nascent peat derived from aquatic vascular plants.

This technique has proved to be central to much of our subsequent work on the fate and transformations of lignocellulose in the Okefenokee. In work conducted to date, we have carried out aerobic degradation studies for periods generally lasting 1 month or less, and anaerobic incubations for up to 1 year. Specific details of recent studies designed to examine the development of lignocellulolytic bacterial populations in acidic Okefenokee Swamp water and sediments are presented in Chapter 3.

Using this approach in combination with standard and modified litter bag incubations in the field, we determined that the rates in aerobic sediments of mineralization of [^{14}C]lignocellulose (labeled either in the lignin or polysaccharide moiety) derived from the dominant Okefenokee sedge, *Carex walteriana,* were two- to threefold lower than rates for structurally similar emergent macrophytes *(C. walteriana, Juncus romerianus, Spartina alterniflora)* under the neutral-pH conditions of salt marsh sediments of Sapelo Island and the Big Cypress Swamp, FL.[13-15] As the component polymers of lignocellulose (cellulose, hemicellulose, and lignin) are major progenitors of peat, the low degradation rates of these polymers appear to be consistent with the greater organic matter (peat) accumulation characteristic of the Okefenokee relative to the salt marsh and circumneutral-pH freshwater wetlands in the same latitude we have examined.

Most, but not all, of the difference in rates of lignocellulosic detrital mineralization between Okefenokee and the other wetland ecosystems examined is attributable to direct effect of pH with only minor differences due to inherent differences in the lignocellulolytic potential (due principally to bacteria) of the resident populations in the various wetlands. When pH of Okefenokee water is experimentally increased to 7, or when bacterial populations from the Okefenokee are collected and resuspended in sterile water from a neutral-pH wetland, rates of lignocellulose degradation supported by the population increase to nearly those of the natural populations from the neutral environments.[16]

VII. IMPLICATIONS FOR ACID INHIBITION OF MICROBIAL PROCESSES IN A DETRITUS-BASED ECOSYSTEM

In that lignocellulose represents the most abundant source of organic carbon in the Okefenokee and that intact lignocellulose is unavailable as a carbon source to most animals, pH limitation of detritus transformation to more utilizable degradation products or highly nutritive bacterial biomass may be important in limiting animal production in the swamp ecosystem. However, this assumption itself supposes that the carbon comprising animal biomass is derived from transformed vascular plant carbon. If, on the other hand, the animal food web is supported by easily utilizable carbon derived from nonlignocellulosic components of higher plants or algae or carbon that is processed microbially by pathways that are not pH sensitive, then reduced rates of lignocellulose transformation might not be closely related to rates of animal production. Recent computer simulation studies based on our accumulated data for microbial and animal production and lignocellulose transformation rates suggest that other sources of organic carbon besides microbial transformation of lignocellulosic detritus are, in fact, needed to support observed animal production rates and standing stocks.[17]

Incubations of natural Okefenokee sediments and microbial assemblages with specifically radiolabeled lignocelluloses reveal also that the lignin and polysaccharide components of a given lignocellulose, although physically bound within the heteropolymer, are transformed by microorganisms at very different rates. For ^{14}C-lignocellulose from *C. walteriana,* for example, the rate of mineralization (to CO_2) of the polysaccharide component was two- to threefold higher than that of the lignin component.[15,16] This difference is consistent for lignocelluloses from all other freshwater, terrestrial, and marine plants we have studied.[12,18,19] The relatively higher rates of mineralization of the polysaccharide components result in the gradual enrichment of remaining detritus in lignin as it is degraded. We have observed such enrichment over 1-year intervals using litter bag incubations as well, thus validating the laboratory studies with labeled lignocellulose (Moran and Hodson, unpublished).

VIII. SUMMARY

Contrary to assumptions made by early Okefenokee investigators, this highly acidic wetland does support abundant and active microbial populations relative even to the most microbiologically rich ecosystems previously studied. The microorganisms rapidly utilize simple dissolved organic compounds and support high rates of secondary production of particulate organic carbon (biomass). No conclusive evidence of reduction in rates of these processes due to the acidic conditions in the Swamp has been observed. In contrast, bacterial isolates from the Okefenokee grow well under rich laboratory conditions at pH 7 but not at the *in situ* pH of 4.

Microorganisms, principally bacteria, are active also in the mineralization of refractory lignocellulosic detritus derived from vascular plants. However, rates of this process appear to be reduced twofold or more by the acidity of the Okefenokee relative to rates in neutral-pH wetland environments. The reduction in lignocellulose mineralization rates is consistent with, but not necessarily a major contributing factor in, the significant accumulation of organic matter as peat in the Okefenokee. Moreover, the rates of conversion of refractory detrital lignocellulose to microbial biomass appear to be insufficient to consider carbon flow from lignocellulose to bacteria to bacterivorous animals to be the major link between primary and secondary production in this ecosystem. Rather we must assume, based on available data, that other, perhaps pH-insensitive carbon flow pathways are equally or more important.

ACKNOWLEDGMENTS

The authors wish to thank Mr. John Schroer, Manager, Okefenokee National Wildlife Refuge, and his staff for continuing assistance with our research in the Okefenokee. Funds for this research were provided by grants BSR 8215587 and BSR 8114823 from the National Science Foundation. Additional support was received from the Athens Environmental Research laboratory, U.S. Environmental Protection Agency.

REFERENCES

1. **Cohen, A. D., Casagrande, D. J., Andrejko, M. J., and Best, G. R.,** Eds., The Okefenokee Swamp: Its Natural History, Geology, and Geochemistry. Wetland Surveys, Inc., Los Alamos, NM, 1984.

2. **Odum, E. P. and de la Cruz, A. A.,** Particulate organic detritus in a Georgia salt marsh-estuarine ecosystem, in *Estuaries*, AAAS Publication 83. Washington, D.C., 1967.

3. **Murray, R. E. and Hodson, R. E.,** Microbial biomass and utilization of dissolved organic matter in the Okefenokee Swamp ecosystem, *Appl. Environ. Microbiol.,* 47, 685, 1984.

4. **Murray, R. E. and Hodson, R. E.,** Annual cycle of bacterial secondary production in five aquatic habitats of the Okefenokee Swamp ecosystem, *Appl. Environ. Microbiol.,* 49, 650, 1985.

5. **Murray, R. E. and Hodson, R. E.,** Influence of macrophyte decomposition on growth rate and community structure of Okefenokee Swamp bacterioplankton, *Appl. Environ. Microbiol.,* 51, 293, 1986.

6. **Benner, R., Maccubbin, A. E., and Hodson, R. E.,** Preparation, characterization, and microbial degradation of specifically radiolabeled ^{14}C-lignocelluloses from marine and freshwater macrophytes, *Appl. Environ. Microbiol.,* 47, 381, 1984.

7. **Latter, P. M., Cragy, J. B., and Heal, O. W.,** Comparative studies on the microbiology of four moorland soils in the Northern Pennines, *J. Ecol.,* 55, 445, 1966.

8. **Collins, V. G., D'Sylva, B. T., and Latter, P. M.,** Microbial populations in peat, in *Production Ecology of Some British Moors and Montane Grasslands,* Heal, O. W. and Perkins, D. F., Eds., Springer-Verlag, New York, 1978, 94.

9. **Martin, N. J. and Holding, A. J.,** Nutrient availability and other factors limiting microbial activity in the blanket peat, in *Production Ecology of Some British Moors and Montane Grasslands,* Heal, O. W. and Perkins, D. F., Eds., Springer-Verlag, New York, 1978, 113.

10. **Moran, M. A., Maccubbin, A. E., Benner, R., and Hodson, R. E.,** Dynamics of microbial biomass and activity in five habitats of the Okefenokee Swamp Ecosystem, *Microb. Ecol.,* 14, 203, 1987.

11. **Crawford, R. L. and Crawford, D. L.,** Radioisotopic methods for the study of lignin biodegradation, *Dev. Ind. Microbiol.,* 19, 35, 1978.

12. **Maccubbin, A. E. and Hodson, R. E.,** Mineralization of detrital lignocelluloses by salt marsh sediment microflora, *Appl. Environ. Microbiol.,* 40, 735, 1980.

13. **Benner, R., Lewis, D. L., and Hodson, R. E.,** Biogeochemical cycling of organic matter in acidic environments: are microbial degradative processes adapted to low pH? in *Acid Stress and Aquatic Microbial Interactions,* Rao, S. S., Ed., CRC Press, Boca Raton, FL, 1989, chap. 3.

14. **Benner, R., Maccubbin, A. E., and Hodson, R. E.,** Anaerobic biodegradation of the lignin and polysaccharide components of lignocellulose and synthetic lignin by sediment microflora, *Appl. Environ. Microbiol.,* 47, 998, 1984.

15. **Benner, R., Maccubbin, A. E., and Hodson, R. E.,** Temporal relationship between the deposition and microbial degradation of lignocellulosic detritus in a Georgia salt marsh and the Okefenokee Swamp, *Microb. Ecol.,* 12, 291, 1986.

16. **Benner, R., Moran, M. A., and Hodson, R. E.,** Effects of pH and plant source on lignocellulosic biodegradation rates in two wetland ecosystems, the Okefenokee Swamp and a Georgia salt marsh, *Limnol. Oceanogr.,* 30, 489, 1985.

17. **Moran, M. A., Legovic, T., Benner, R., and Hodson, R. E.,** Carbon flow from lignocellulose: a simulation analysis of a detritus-based ecosystem, *Ecology,* 69, 1525, 1988.

18. **Benner, R. and Hodson, R. E.,** Microbial degradation of the leachable and lignocellulosic components of leaves and wood from *Rhizophora mangle* in a tropical mangrove swamp, *Mar. Ecol. Prog. Ser.,* 23, 221, 1985.

19. **Hodson, R. E. and Benner, R.,** Anaerobic biodegradation of natural lignocelluloses and synthetic lignin, in Biodeterioration 6. Barry, S. and Houghton, D. R., Eds., CAB International Mycological Institute, Slough, England.

20. **Erkenbrecher, C. W. and Stevenson, L. H.,** *Limnol. Oceanogr.,* 20, 618, 1975.

21. **Axelrod, D. M., Moore, K. A., and Bender, M. E.,** Virginia Water Res. Center Bull. 79, 1976.

22. **Hanson, R. B. and Snyder, J. T.,** *Limnol. Oceanogr.,* 60, 99, 1979.

23. **Murrary, R. and Hodson, R. E.,** *Appl. Environ. Microbiol.,* 47, 685, 1984.

24. **Yingst, J. V.,** *Mar. Biol.,* 47, 41, 1978.

25. **Christian, R. R., Bancroft, K., and Wiebe, W. J.,** *Soil Sci.,* 119, 89, 1975.

26. **Jenkinson, D. S. and Oades, J. M.,** *Soil Biol. Biochem.,* 11, 193, 1979.

27. **Ross, D. J., Tate, K. R., Cairnes, A., and Pansier, E. A.,** *Soil Biol. Biochem.,* 12, 375, 1980.

28. **Cunningham, H. W. and Wetzel, R. G.,** *Limnol. Oceanogr.,* 23, 166, 1978.

Chapter 5

SULFUR BIOGEOCHEMISTRY OF AN ACIDIC LAKE IN THE ADIRONDACK REGION OF NEW YORK

Jeffrey S. Owen and Myron J. Mitchell

TABLE OF CONTENTS

I. INTRODUCTION

Understanding biogeochemical cycling of sulfur in limnetic systems necessitates quantifying the dynamics of inorganic and organic sulfur chemical species in water, seston, and sediment. It has been shown that sulfate plays a critical role in regulating mineral acidity in surface waters and there has been increasing effort on quantifying the flux of this anion within watersheds, some of which include lakes.[1-5] For soils within a watershed, sulfate has a major role in regulating the removal of nutrient cations and alkalinity due its association with the hydronium ion and its importance as a counter-ion.[6-7] These watershed processes contribute to the chemistry of surface waters by regulating sulfur transport from the watershed to lake systems. Of special importance is the role of sulfur transformations in lakes affecting alkalinity generation.[8-13] It has been shown that dissimilatory reduction of sulfate within lakes can consume substantial quantities of hydronium ions.[4,8,9,11-14]

Previous studies have relied on a variety of techniques in determining various aspects of limnetic sulfur budgets including input-output watershed budgets,[1-3] mass balance approaches using sediment traps and net sediment accumulation rates,[5,15] pore water chemistry,[11,12,16] stable isotopes,[17,18] and radioisotopes.[10] The objectives of this chapter are to (1) present new information on pore water chemistry and sediment sulfur speciation from acidic South Lake and (2) combine these findings with previous data on the sulfur biogeochemistry of this system. This approach allows us to compare, using independent estimates, sulfur incorporation processes within South Lake and to ascertain the relative importance of seston accumulation and pore water sulfate flux to total sulfur incorporation in sediment.

II. METHODS

A. Site Description

South Lake is located in the southwestern Adirondack Mountains, NY and has been the site of previous investigations of sulfur biogeochemistry.[19,20] Basic limnological information for South Lake is given in Table 1. South Lake is oligotrophic and has lost most of its fish population.[21] Bathymetry and the location of the sediment core, water column, and pore water sampling site are shown in Figure 1.

B. Sampling and Analytical Techniques

A sediment core was taken using a Kajak-Brinkhurst gravity corer. The core was 4.7 cm in diameter and was sectioned into 1-cm intervals to a depth of 24 cm and then into 2-cm intervals until 30-cm depth. Sectioning was performed inside a glove box with a nitrogen atmosphere to minimize oxidation of reduced chemical species. Sediment samples were then placed into plastic bags and stored moist at 1°C in a nitrogen-filled container until analysis.

Various sulfur factions were analyzed by conversion of an operationally defined fraction to sulfide using a modified Johnson-Nishita apparatus.[22,23] The detection limit for the colorimetric determination of sulfide was found to be 1 μg.

Acid-volatile sulfur[24] was recovered by addition of 1:1 HCl. The acid-volatile fraction includes H_2S and nonpyritic metallic sulfide compounds. This procedure has been shown to not reduce pyritic or organic sulfur.[25]

Elemental sulfur was determined on selected subsamples by extraction of wet sediment in 20 ml of acetone.[25] Elemental sulfur was extracted from wet sediment for 16 h using a wrist action shaker. This mixture was filtered, rinsed in acetone, and analyzed for chromium-reducible sulfur.

Chromium-reducible sulfur (pyritic and elemental sulfur) was recovered by addition of Cr(II) solution and concentrated HCl.[26] The specificity of this technique has not been adequately addressed.[27] Some researchers[28,29] have used the Cr(II) method to sequentially

Table 1
LIMNOLOGICAL
CHARACTERISTICS OF SOUTH
LAKE

Location	43° 30′ N 74° 52′ W
Elevation (m)	615
Lake area (ha)	202
Watershed area (ha)	1420
Flushing rate (year^{-1})	0.5
Mean depth (m)	13
Maximum depth (m)	26
pH	4.9 — 5.4
Acid-neutralizing ca- pacity (μeq l^{-1})	−3 — 10

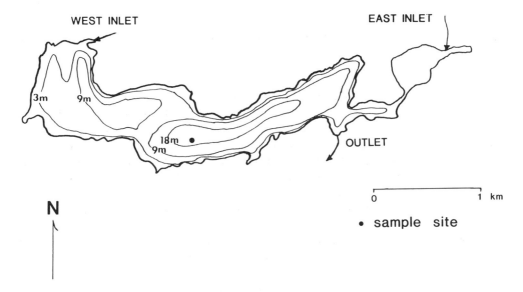

FIGURE 1.   Bathymetry of South Lake. Sampling site was at 18-m depth.

extract only pyritic and elemental sulfur, but Brown[27] suggested the procedure reduced some humic sulfur in organic-rich samples. Due to the uncertain nature of Cr(II)-reducible sulfur, we will term this fraction "chromium-reducible sulfur". Chromium-reducible sulfur values presented here have been corrected for acid-volatile and elemental sulfur.

Hydriodic acid (HI)-reducible sulfur was recovered by addition of a mixed reagent (HI) of hydriodic, formic, and hypophosphorous acids.[30] This fraction consisted of nonpyritic inorganic sulfur and ester sulfate.

Carbon-bonded sulfur was determined by subtracting HI-reducible sulfur and chromium-reducible sulfur from total sulfur. Due to the uncertainty in the chromium-reducible sulfur fraction, this fraction represented a minimum estimate of carbon-bonded sulfur and was termed "residual sulfur". Total sulfur was quantified on dry and pulverized sediment using an alkaline oxidation procedure[31] as modified by Landers et al.[23]

Sediment moisture content was determined on homogenized subsamples by drying to constant mass at 90°C. A portion of the dry sediment was placed in a crucible and combusted at 550°C for 3 h to determine loss on ignition. Total carbon and nitrogen were determined on 10- to 20-mg subsamples of dry sediment using a Perkin-Elmer 240 C Elemental Analyzer. The instrument was calibrated daily using acetanilide standards. Detection limits were found to be 5 μg nitrogen and 3 μg carbon.

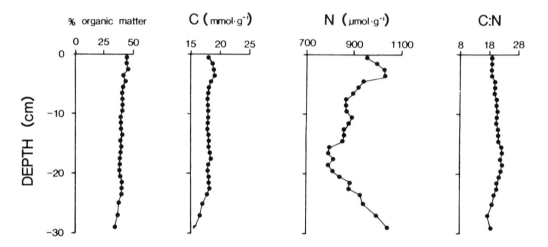

FIGURE 2.   Distribution of organic matter, total carbon, total nitrogen, and carbon/nitrogen ratio in South Lake sediment.

The pore water equilibrator (peeper) used at South Lake consisted of a Plexiglas frame with holes to contain modified polycarbonate centrifuge tubes.[16] The tubes had rectangular holes cut out on two sides and covered with 0.2-μm pore size dialysis membrane. The tubes were completely filled with deionized and deoxygenated water and sealed with rubber septum caps. The peeper was installed by scuba divers on June 19, 1986 at a depth of 18 m. The peeper was held in place at the sediment-water interface by a horizontal Plexiglas support. After 28 d the peeper was retrieved. Samples for determination of pH, sulfate, and nitrate were transferred to samples vials and packed in ice for transport to the laboratory. pH was measured potentiometrically using a combination microprobe. Sulfate and nitrate were analyzed by ion chromatography.

Dissolved oxygen in the water column was determined on a subsample that was fixed immediately following collection and analyzed using the Winkler technique.[32]

III. RESULTS AND DISCUSSION

A. Sediment Organic Matter

South Lake was thermally stratified at the time of sampling and the hypolimnetic water was at 60% O_2 saturation. *In situ* temperature of the hypolimnion and the surficial (0 to 4 cm) sediment was 9°C. Sediment from South Lake was highly flocculent and moisture content was greater than 90% wet mass throughout the core. Organic matter content decreased from 45% dry mass in surficial sediment to 40% at 30 cm depth (Figure 2). ^{210}Pb dating of an adjacent core showed that 10 cm of sediment accumulated since approximately 1916.[15]

The total carbon profile was similar to the organic matter profile (Figure 2). A close relationship between sediment total carbon and organic matter content has been documented for several lakes at the Experimental Lake Area, Ontario, and is probably found in many oligotrophic lakes.[33]

The decomposition of nitrogen-containing compounds is the most important process affecting the nitrogen content of lake sediments. Keeney et al.[34] found 98% of the nitrogen present as organic nitrogen in sediment from a variety of lakes. The decrease in nitrogen concentration with sediment depth below a subsurface maximum (Figure 2) may indicate either organic matter sedimentation has changed with time or that nitrogenous compounds were preferentially mineralized relative to other organic compounds during organic matter decomposition in South Lake. The slight increase in the sediment carbon/nitrogen ratio with sediment depth until about 20 cm reflected this decrease in nitrogen concentration. Thus, South Lake sediment at this hypolimnetic site may be characterized as rich in organic matter.

Previous work by David and Mitchell[5] indicated that the organic content of South Lake seston did not decrease with depth in the water column. They attributed the absence of greater carbon mineralization with depth to low rates of decomposition, possibly related to the cold and acidic condition of the water column. The seston deposited at the South Lake sediment-water interface has likely undergone relatively little oxidation, despite the oxic condition. Landers and Mitchell[10] showed that South Lake had the lowest sediment respiration rate (50 mmol O_2 m^{-2} d^{-1}) of three lakes in laboratory incubation experiments. Rudd et al.[11] found microbial activity and alkalinity generation (via dissimilatory sulfate reduction and denitrification) lower in organic-rich sediments than in highly inorganic sediments. Differences in sediment type and microbial activity could be related to differences in rates of *in situ* lake alkalinity generation.

B. Distribution of Sediment Sulfur Constituents

The total sulfur profile (Figure 3) was similar to a previous sulfur profile found in South Lake sediment.[15] A decrease in total sulfur concentration with sediment depth below a subsurface maximum has been shown for several Adirondack lakes.[35]

Acid-volatile sulfur represented less than 10% of total sulfur throughout the sediment profile and varied little with depth (Figure 3). Elemental sulfur was also a minor constituent, representing less than 5% of total sulfur (not shown). The fraction of total sulfur as residual sulfur varied but generally increased with sediment depth to 70% at 30 cm (Figure 3). HI-reducible sulfur (ester sulfate) was greater than 40% of total sulfur in surficial (0 to 4 cm) sediment and decreased below this depth to about 25% of total sulfur (Figure 3).

The fraction of total sulfur as chromium-reducible sulfur varied with sediment depth (Figure 3). An increase in chromium-reducible sulfur was indicated between 3- and 9-cm depth. Chromium-reducible sulfur became less than 10% of total sulfur below 22-cm depth.

There are a variety of mechanisms for sulfur retention in lake sediments.[35] An increase in limnetic sulfate concentration will increase the rate of dissimilatory sulfate reduction in anaerobic water or sediments.[4] Assimilatory sulfate incorporation (ester sulfate formation) or assimilatory sulfate reduction (carbon-bonded sulfur formation) can also generate alkalinity.[10] An increase in limnetic sulfate concentration will enhance rates of these processes.[35] To ascertain the relative importance of each pathway to total sulfur accumulation, we will compare the net accumulation rate of sulfur inputs from seston deposition and porewater sulfate flux.

The predominance of organic sulfur in lake sediments has been demonstrated by previous studies.[36,37] "Organic sulfur" includes the residual (carbon-bonded) and HI-reducible sulfur (ester sulfate) pools analyzed in this study. These two types of organic sulfur are important to distinguish on the basis of their formation and because ester sulfate represents an oxidized sulfur form while carbon-bonded sulfur is in a reduced state. Formation of ester sulfate does not involve reduction of sulfate but consumes 1 mol of hydronium ion for each mole of sulfate incorporated. Formation of carbon-bonded sulfur consumes 2 mol of hydronium ion for each mole of sulfate reduced.[10,12] Carbon-bonded sulfur may accumulate from deposited seston or be formed *in situ* during organic matter diagenesis.[17]

In the present study, formation of carbon-bonded sulfur during sediment diagenesis might be inferred from the increase in the residual sulfur fraction with depth. Nriagu and Soon[17] showed evidence for *in situ* formation of organic sulfur in lake sediments. Their data suggested that a large portion of sediment organic sulfur was formed by reaction of organic matter with reduced sulfur species such as H_2S and polysulfides. The mechanisms of these reactions are poorly understood and their occurrence in freshwater lake sediments was heretofore unknown.

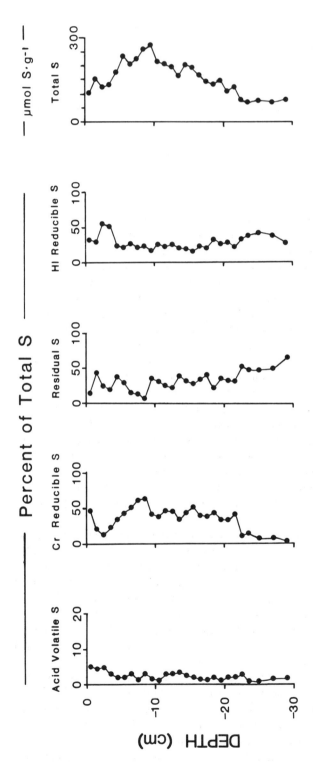

FIGURE 3.   Distribution of sulfur constituents in South Lake sediment. Note that total sulfur is expressed in micromole sulfur per gram.

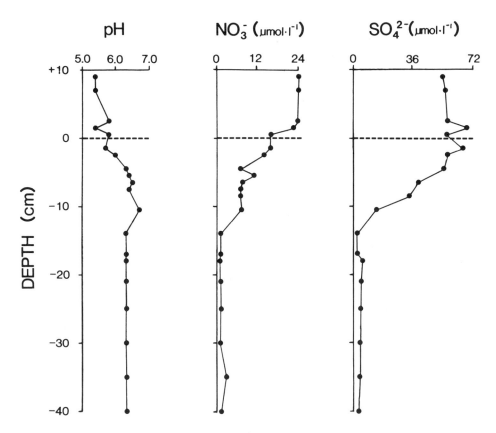

FIGURE 4. Distribution of pH, NO_3^-, and SO_4^{2-} in South Lake pore water. Dashed line indicates sediment-water interface.

Sulfur incorporation via reactions between organic matter and reduced sulfur species has been studied in marine sediments. Casagrande et al.[38] showed that H_2S reacted with organic matter in peat to form organic sulfur compounds. Nissenbaum and Kaplan[39] suggested that sulfur in marine humic material was introduced during *in situ* diagenesis. Brassell et al.[40] identified alkyl thiophenes from the hydrocarbon fraction of marine sediments. The origin of these thiophenes was low temperature incorporation of reduced sulfur into lipid molecules. Mango[41] indicated that diagenetic sulfur enrichment of carbohydrates was base catalyzed. Recent work by Francois[42,43] showed that such reactions occur, albeit slowly, at "seawater pH (8.2)". In freshwater lake sediments (circumneutral pH), carbon/sulfur ratios and stable sulfur isotope data suggest that reactions between reduced sulfur and organic matter are important.[17]

The conversion of HI-reducible sulfur (ester sulfate) to reduced sulfur has been demonstrated in peat and lake sediments.[17,45] In South Lake the decrease in HI-reducible sulfur below 4-cm depth (Figure 3) indicated the labile nature of ester sulfate in sediment. Mineralization of ester sulfate can be accomplished by the action of sulfatases, activity of which generally decreases with sediment depth.[17,37] David and Mitchell[5] found HI-reducible sulfur was the more labile organic sulfur constituent in South Lake sediment. Conversion of HI-reducible sulfur into residual or chromium-reducible forms may be indicated by the increase with sediment depth of chromium-reducible sulfur (below 3 cm) and residual sulfur (below 10 cm).

C. Ion Fluxes Across the Sediment-Water Interface

The pH, sulfate, and nitrate profiles in South Lake pore water are shown in Figure 4. We calculated concentration gradients of H^+, SO_4^{2-}, and NO_3^- and estimated the net flux using

Table 2
ION FLUXES IN SOUTH LAKE SEDIMENT

	Porosity (%)	Gradient (μmol cm^{-3} cm^{-1})	Diffusion coefficient (cm^2 s^{-1})	Flux (μmol m^{-2} d^{-1})
H$^+$	96	-2.53×10^{-4}	7.45×10^{-5}	2.0
SO$_4$$^{2-}$	96	-4.80×10^{-3}	6.86×10^{-6}	27.0
NO$_3$$^-$	96	-1.00×10^{-3}	1.30×10^{5}	1.0

Fick's first law of diffusion: $J = -D\emptyset dC/dX$ where J is the flux in concentration per unit area per unit time, D is the diffusion coefficient in square centimeters per second, $\emptyset$ is the sediment porosity, and dC/dX is the concentration gradient.[46] Diffusion coefficients at *in situ* temperature were calculated from the diffusion coefficients at infinite dilution at 0°C using the equations given by Li and Gregory.[47]

Table 2 shows the ion fluxes in μmol m^{-2} d^{-1}. Since a negative gradient indicated a lower concentration at depth in the sediment, a positive flux indicated ion movement from the water to the sediment (incorporation). If we assumed that annual flux is similar to that observed during summer stratification, the estimated NO$_3$$^-$ loss to the sediments was 3.9 mmol nitrogen m^{-2} year^{-1}. Nitrate was not removed (<2 μmol l^{-1}) until below 13 cm sediment depth. This contrasted with data derived from highly inorganic epilimnetic sediments in several acidic lakes where NO$_3$$^-$ consumption was rapid and resulted in depletion of pore water NO$_3$$^-$ within a few centimeters of the sediment-water interface.[12]

The annual H$^+$ flux in South Lake sediment was 5.7 mmol hydrogen m^{-2} year^{-1}. This estimate was similar to those for hypolimnetic sediments in Big Moose Lake,[48] but lower than that reported for other acid lakes.[12,49] The flux of H$^+$ between sediment and overlying water is a function of alkalinity-generating processes such as dissimilatory sulfate reduction and denitrification, which affect the pore water SO$_4$$^{2-}$ and NO$_3$$^-$ gradients.

The estimated SO$_4$$^{2-}$ flux to South Lake sediment was 10.0 mmol m^{-2} year^{-1}. Carignan[49] calculated a mean SO$_4$$^{2-}$ flux of 119 mmol m^{-2} year^{-1} based on pore water gradients at three sites in acidic Clearwater Lake. Cook et al.[14] and Rudd et al.[12] reported SO$_4$$^{2-}$ fluxes of 18, 40, and 24 mmol m^{-2} year^{-1} for Little Rock Lake, Lake 233, and Dart's Lake, respectively. Our measured flux (incorporation) rates indicate that microbially mediated processes in South Lake are less important than for some freshwater lake sediments.

D. Processes Regulating Sulfur Incorporation into Sediment

The solid phase sediment data from South Lake indicated that organic sulfur (HI-reducible and residual sulfur) were the major sulfur constituents in the surficial sediment. A large portion of this organic sulfur may be derived from seston.[5] Accumulation of seston might be an important pathway for sulfur incorporation because seston has been shown to contain approximately 0.9% sulfur in a variety of aquatic systems.[4,5,17,50] David and Mitchell[5] presented a sulfur budget for South Lake based on mass balance estimates that showed 19.2 mmol sulfur m^{-2} year^{-1} were derived from seston. Based on their estimated seston mineralization rate of 26%, net seston sulfur accumulation was 14.4 mmol sulfur m^{-2} year^{-1}.

In the present study, the estimated pore water SO$_4$$^{2-}$ incorporation rate was 10.0 mmol sulfur m^{-2} year^{-1}. Landers and Mitchell[10] showed that 14.7 mmol sulfur m^{-2} year^{-1} were added by SO$_4$$^{2-}$ incorporation from overlying water in incubated sediment cores from South Lake. These two independent estimates of SO$_4$$^{2-}$ incorporation rates are similar.

The data indicate that seston inputs have a similar contribution as SO$_4$$^{2-}$ incorporation processes to sulfur retention in organic-rich South Lake sediment. Attempts to model sulfur incorporation in sediments that underestimate the seston contribution might erroneously suggest an unimportant role for this mechanism of sulfur retention.[16,18] It is therefore important to measure the contribution of both seston and SO_4^{2-} incorporation to sulfur retention on limnetic systems with low rates of microbial activity (sulfate reduction) in the sediments.

ACKNOWLEDGMENTS

We would like to express our gratitude to G. R. Holdren for use of the peeper and to S. C. Schindler for installation. The comments of C. A. Kelly were greatly appreciated. This research was supported in part by the Electric Power Research Institute, Palo Alto, CA.

REFERENCES

1. **Galloway, J. N., Schofield, C. L., Hendrey, G. R., Altwicker, E. R., and Troutman, D. E.,** Sources of acidity in three lakes during snowmelt, *Proc. Int. Conf. Ecological Impact of Acid Precipitation,* SNSF project, Oslo, Norway, 1980, 264.
2. **Galloway, J. N., Schofield, C. L., Peters, N. E., Henry, G. R., and Altwicker, E. R.,** Effect of atmospheric sulfur on the composition of three Adirondack lakes, *Can. J. Fish. Aquat. Sci.,* 40, 799, 1983.
3. **Wright, R. F.,** Input-output budgets in Langtjern, a small acidified lake in southern Norway, *Hydrobiologia,* 101, 1, 1983.
4. **Cook, R. B. and Schindler, D. W.,** Biogeochemistry of sulfur in an experimentally acidified lake, *Ecol. Bull.,* 35, 115, 1983.
5. **David, M. B. and Mitchell, M. J.,** Sulfur constituents and cycling in waters, seston and sediment of an oligotrophic lake, *Limnol. Oceanogr.,* 30, 1196, 1985.
6. **Reuss, J. O. and Johnson, D. W.,** *Acid Deposition and the Acidification of Soils and Waters,* Springer-Verlag, New York, 1986, 119.
7. **Mitchell, M. J. and Fuller, R. D.,** Models of sulfur dynamics in forest and grassland ecosystems with emphasis on soil processes, *Biogeochemistry,* 5, 133, 1988.
8. **Kelly, C. A., Rudd, J. W. M., Cook, R. B., and Schindler, D. W.,** The potential importance of bacterial processes in regulating the rate of lake acidification, *Limnol. Oceanogr.,* 27, 868, 1982.
9. **Kelly, C. A. and Rudd, J. W. M.,** Epilimnetic sulfate reduction and its relationship to lake acidification, *Biogeochemistry,* 1, 63, 1984.
10. **Landers, D. H. and Mitchell, M. J.,** ^{35}Sulfate incorporation in the sediments of three New York lakes, *Hydrobiologia,* 160, 85, 1988.
11. **Rudd, J. W. M., Kelly, C. A., and Furutani, A.,** The role of sulfate reduction in long term accumulation of organic and inorganic sulfur in lake sediments, *Limnol. Oceanogr.,* 31, 1281, 1986.
12. **Rudd, J. W. M., Kelly, C. A., St. Louis, V., Hesslein, R. H., Furutani, A., and Holoka, M. H.,** Microbial consumption of nitric and sulfuric acids in acidified north temperate lakes, *Limnol. Oceanogr.,* 31, 1267, 1986.
13. **Schafran, G. C. and Driscoll, C. T.,** Comparison of terrestrial hypolimnetic sediment generation of acid neutralizing capacity for an acidic Adirondack lake, *Environ. Sci. Technol.,* 21, 988, 1987.
14. **Cook, R. B., Kelly, C. A., Kingston, J. C., and Kreis, R. G.,** Chemical limnology of soft water lakes in the upper Midwest, *Biogeochemistry,* 4, 97, 1987.
15. **Mitchell, M. J., David, M. B., and Uutala, A. J.,** Sulfur distribution in lake sediment profiles as an index of historical depositional patterns, *Hydrobiologia,* 121, 121, 1985.
16. **Holdren, G. R., Jr., Brunelle, T. M., Matisoff, G., and Whalen, M.,** Timing the increase in atmospheric sulphur deposition in the Adirondack Mountains, *Nature,* 311, 245, 1984.
17. **Nriagu, J. O. and Soon, Y. K.,** Distribution and isotopic composition of sulfur in lake sediments of northern Ontario, *Geochim. Cosmochim. Acta,* 49, 823, 1985.
18. **Fry, B.,** Stable sulfur isotopic distribution and sulfate reduction in lake sediments of the Adirondack Mountains, New York, *Biogeochemistry,* 2, 329, 1986.
19. **Mitchell, M. J., Landers, D. H., and Brodowski, D. F.,** Sulfur constituents of sediments and their relationship to lake acidification, *Water Air Soil Pollut.,* 16, 177, 1981.
20. **Mitchell, M. J., Landers, D. H., Brodowski, D. F., Lawrence, G. B., and David, M. B.,** Organic and inorganic sulfur constituents of the sediments in three New York lakes: effect of site, sediment depth and season, *Water Air Soil Pollut.,* 21, 231, 1984.

21. **Pfeiffer, M. H. and Festa, P. J.**, Acidity Status of Lakes in the Adirondack Region of New York in Relation to Fish Resources, New York State Department of Conservation, Albany, New York, 1980.
22. **Johnson, C. M. and Nishita, H.**, Microestimation of sulfur in plant materials, soils, and irrigation waters, *Anal. Chem.*, 24, 736, 1952.
23. **Landers, D. H., David, M. B., and Mitchell, M. J.**, Analysis of organic and inorganic sulfur constituents in sediments, soil, and water, *Int. J. Environ. Anal. Chem.*, 14, 245, 1983.
24. **Smittenberg, J., Harmsen, G. W., Quispel, A., and Otzen, D.**, Rapid methods for determining different types of sulphur compounds in soils, *Plant Soil*, 3, 353, 1951.
25. **Wieder, R. K., Lang, G. E., and Granus, V. A.**, An evaluation of wet chemical methods for quantifying sulfur fractions in freshwater wetland peat, *Limnol. Oceanogr.*, 30, 1109, 1985.
26. **Zhabina, N. N. and Volkov, I. I.**, A method of determination of various sulfur compounds in sea sediments and rocks, in *Environmental Biogeochemistry and Geomicrobiology*, Vol. 3, Krumbein, W. E., Ed., Ann Arbor Science, Ann Arbor, MI, 1978, 735.
27. **Brown, K. A.**, Formation of organic sulphur in anaerobic peat, *Soil Biol. Biochem.*, 18, 131, 1986.
28. **Westrich, J. T.**, The Consequences and Controls of Bacterial Sulfate Reduction in Marine Sediments, Ph.D. thesis, Yale University, New Haven, CT, 1983.
29. **Howarth, R. W. and Merkel, S.**, Pyrite formation and the measurement of sulfate reduction in salt marsh sediments, *Limnol. Oceanogr.*, 29, 598, 1984.
30. **Johnson, C. M. and Ulrich, A.**, Analytical methods for use in plant analysis, *Calif. Agric. Exp. St. Bull.*, 766, 25, 1959.
31. **Tabatabai, M. A. and Bremner, J. M.**, An alkaline oxidation method for determination of total sulfur in soils, *Soil Sci. Soc. Am. Proc.*, 34, 62, 1970.
32. *Standard Methods for the Examination of Water and Wastewater*, 15th ed., American Public Health Association, New York, 1980.
33. **Brunskill, C. J., Povoledo, D., Graham, B. W., and Stainton, M. P.**, Chemistry of surface sediments of sixteen lakes in the Experimental Lakes Area, northwestern Ontario, *J. Fish. Res. Board Can.*, 28, 277, 1971.
34. **Keeney, D. R., Konrad, J. G., and Chesters, G.**, Nitrogen distribution in some Wisconsin lake sediments, *J. Water Pollut. Control. Fed.*, 42, 411, 1970.
35. **Mitchell, M. J., Schindler, S. C., Owen, J. S., and Norton, S. A.**, Comparison of sulfur concentrations within lake sediment profiles, *Hydrobiologia*, 157, 219, 1988.
36. **Hesse, P. R.**, The distribution of sulphur in the muds, water, and vegetation of Lake Victoria, *Hydrobiologia*, 11, 29, 1958.
37. **King, G. M. and Klug, M. J.**, Sulfhydrolase activity in sediments of Wintergreen Lake, Kalamazoo County, Michigan, *Appl. Environ. Microbiol.*, 39, 950, 1980.
38. **Casagrande, D. J., Idowu, G., Friedman, A., Rickert, P., Siefert, K., and Schlenz, D.**, H_2S incorporation in coal precursors: origins of organic sulphur in coal, *Nature*, 282, 599, 1979.
39. **Nissenbaum, A. and Kaplan, I. R.**, Chemical and isotopic evidence for the in situ origin of marine humic substances, *Limnol. Oceanogr.*, 17, 570, 1972.
40. **Brassell, S. C., Lewis, C. A., deLeeuw, J. W., deLange, F., and Sinninghe Damste, J. S.**, Isoprenoid thiophenes: novel products of sediment diagenesis?, *Nature*, 320, 160, 1986.
41. **Mango, F. D.**, The diagenesis of carbohydrates by hydrogen sulfide, *Geochim. Cosmochim. Acta*, 47, 1433, 1983.
42. **Francois, R.**, A study of sulphur enrichment in the humic fraction of marine sediments during early diagenesis, *Geochim. Cosmochim. Acta*, 51, 17, 1987.
43. **Francois, R.**, A study of the extraction conditions of sedimentary humic acids to estimate their true in situ sulfur content, *Limnol. Oceanogr.*, 32, 964, 1987.
44. **Howarth, R. W.**, The ecological significance of sulfur in the energy dynamics of salt marsh and coastal marine sediments, *Biogeochemistry*, 3, 5, 1984.
45. **Altschuer, Z. S., Schnepfe, M. M., Sibler, C. C., and Simon, F. O.**, Sulfur diagenesis in Everglades peat and origin of pyrite in coal, *Science*, 221, 221, 1983.
46. **Berner, R. A.**, *Early Diagenesis*, Princeton University Press, Princeton, NJ, 1980.
47. **Li, Y. and Gregory, S.**, Diffusion of ions in sea water and in deep-sea sediments, *Geochim. Cosmochim. Acta*, 38, 703, 1974.
48. **Owen, J. S.**, Sediment Sulfur Biogeochemistry in Two Acidic Lakes, M.S. thesis, State University of New York College of Environmental Science and Forestry, Syracuse, 1986.
49. **Carignan, R.**, Quantitative importance of alkalinity flux from the sediments of acid lakes, *Nature*, 317, 158, 1985.
50. **Cuhel, R. L., Taylor, C. D., and Jannasch, H. W.**, Assimilatory sulfur metabolism in marine microorganisms: considerations for the application of sulfate incorporation into protein as a measurement of natural population protein synthesis, *Appl. Environ. Microbiol.*, 43, 160, 1982.

Chapter 6

EPILITHIC MICROBIAL POPULATIONS AND LEAF DECOMPOSITION IN ACID-STRESSED STREAMS

Anthony V. Palumbo, Patrick J. Mulholland, and Jerry W. Elwood

TABLE OF CONTENTS

ABSTRACT

The number of bacteria on decomposing leaf material (direct counts of acridine orange-stained bacteria) and bacterial production on leaf material and on rock surfaces (incorporation of tritiated thymidine into dioxyribonucleic acid, DNA) were measured as part of a study of the effects of acidification on microbial communities in streams. The number of bacteria on leaf material ranged from 3.3×10^{10} to 12.4×10^{10} cells per gram dry weight, it was not significantly different among streams of different pH and aluminum concentration, it increased over the 1st 4 weeks of incubation, and it was relatively constant therafter at all sites. There was no significant dilution of added thymidine by unlabeled internal or external pools of thymidine in measurements of bacterial production on leaf material. Bacterial production on both leaf and rock surfaces was significantly greater in streams with higher pH (>5.5) than at matched sites in nearby streams with lower pH (<5.0). However, differences in productivity among sites were not evident until after 2 weeks of incubation of leaf material in the streams, and significant differences could not be generated in short-term (<48-h) transplants of rock material from high to low pH sites. Therefore, the effects of acidification on bacteria do not appear to be due to acute toxicity. Because Al concentration and pH covary in these streams (high Al concentrations at low pH), the observed depression of bacterial production at lower pH sites may be due to chronic effects of high concentrations of both H^+ and Al^{3+} or may be indirectly due to the effects of acidification on the invertebrate grazing community in the streams.

I. INTRODUCTION

Bacteria on rocks and decomposing leaves are significant components of the total microbial community in streams. These communities are the most important bacterial communities in the high elevation streams in the Great Smoky Mountains National Park (GSMNP) due to the lack of fine grain sediments and the relatively low activity and biomass of the bacterioplankton.[1] Thus, knowledge of the productivity of bacteria on rock and leaf surfaces is important in assessing the effects of acidification on microbial communities in these streams.

Although decomposition of leaf material in aquatic systems has been extensively studied, there is very little information on the production of bacteria on decomposing leaf material. Numbers of bacteria,[2] adenosine triphosphate (ATP) content,[3] fungal biomass,[4,5] microbial respiration,[6] and numerous nutrient parameters[7] have been measured, but technique limitations have until recently prevented investigators from measuring bacterial production.

The incorporation of tritiated thymidine into DNA has become a widely used method for measuring bacterial production in the water column[8] and in aquatic sediments,[9,10] and has recently been applied to bacteria on leaf material.[11] Simultaneous measurements of numbers of bacteria and the rate of thymidine incorporation into DNA permits calculation of specific bacterial productivity (per cell).

Acidification appears to reduce decomposition rates in a variety of aquatic systems[2,12,13] including streams.[14,15] In streams, this effect may be due to direct effects on the bacteria, on insects, or on both. Leaf-shredding insects are important in the breakdown of leaf material.[16-19] In a number of studies, reduced numbers of invertebrates were found associated with decomposing leaf material in acidic systems, and much of the reduction in leaf weight loss was attributed to reduced rate of shredding by macroinvertebrates.[14,20] However, the effects of acidification on the microbial community may also play an important role in reducing decomposition rate. Decomposing plant material has been found to have lower numbers of bacteria and reduced bacterial activity associated with it under acidic conditions[21-24] than under conditions of higher pH. In an earlier study,[11] we found that, during decomposition of leaf material, tritiated thymidine incorporation (an indicator of bacterial production) was significantly reduced under acidic conditions even when macroinvertebrates were excluded. However, only two sites were used in that study and the higher pH site also had higher P levels; thus, the differences may have been related to P rather than to acidification. Also, the conditions were somewhat artificial because the leaf material was incubated in chambers to eliminate possible effects of differences in the macroinvertebrate community between sites.

Data on epilithic bacteria in streams are limited. Numbers of bacteria[25] and bacterial activity[26] have been measured, and ATP has been used as an indicator of total microbial biomass[1,27] on rocks. Also, the interactions between bacteria and algae in the epilithic community[28] and the utilization of riverborne organic carbon by epilithic bacteria have been examined.[29] Only recently have tritiated thymidine techniques been applied to assess epilithic bacterial production.[1] Acidification effects on epilithic communities are not well documented, although there appear to be significant effects of acidification on epilithic bacterial communities. Both ATP and tritiated thymidine incorporation into DNA[1] have been found to be higher at sites with pH >5.8 than at similar sites with pH <5.0, and there is evidence for elimination of some types of bacteria at low pH.[27]

The purpose of this study was to determine the rate of bacterial production on decomposing leaf material and on rock surfaces in streams with a range of pH and Al levels. Differences in bacterial production on leaf material were further related to reductions in the bacterial population size or to per cell productivity and to rates of leaf litter decomposition and Al accumulation.[30]

II. METHODS

A. Study Sites

The study was conducted in 1984 and 1985, primarily in the Walker Camp Prong watershed, GSMNP, near the Tennessee-North Carolina border (Figure 1). The four main study sites [Cole Creek (CO), Trout Branch (TB), Walker Prong 4 (WP-4), and Walker Prong 5 (WP-5)] are all second-order streams of similar size but differing baseflow pH (4.5 to 6.4). Streams in the Walker Camp watershed are acidified to different degrees, presumably due to differences in the amount of exposed pyritic rock material in subcatchments. Outcrops of pyritic phyllite are subject to oxidation and reduce the pH in the streams.[31]

The four study sites consist of two sets of matched streams (CO/TB and WP-4/WP-5) based on proximity, stream size, order, and elevation, and were used for studies of bacteria on leaf material and rocks. CO and TB are tributaries of Walker Camp Prong that drain the south slope of Mt. LaConte and are within 1 km of each other. CO is the less acidic of the sites; it has a higher baseflow pH and a higher acid-neutralizing capacity (ANC) and lower concentrations of total monomeric Al (Table 1). The other two main sites, WP-4 and WP-5, are at higher elevation. WP-5 is the less acidic site, with higher ANC and lower concentrations of total monomeric Al (Table 1). Both total soluble P (TSP) and soluble reactive P (SRP) are very low at all sites, but are slightly higher at CO than at TB and are higher at WP-4 than at WP-5 (Table 1). In the more acidic sites (WP-4 and TB), concentrations of manganese and, to a lesser extent, iron are slightly elevated above levels at the less acidic sites.[32] Concentrations of other trace metals, such as Hg, Cd, Pb, Cu, and Zn, commonly observed in streams affected by acid mine drainage were undetectable. Additional chemistry data for these streams are given by Mulholland et al.[32] and Palumbo et al.[11]

Vegetation in the Walker Camp Prong watershed is dominated by mature stands of red spruce, Fraser fir, and yellow birch at the higher elevations; beech and hemlock become important at lower elevations and in riparian zones. A subcanopy is formed by dense stands of *Rhododendron* at high elevations and near the streams.

Two additional pairs of sites, Walker Camp Prong 1 (WP-1)/Walker Camp Prong 3 (WP-

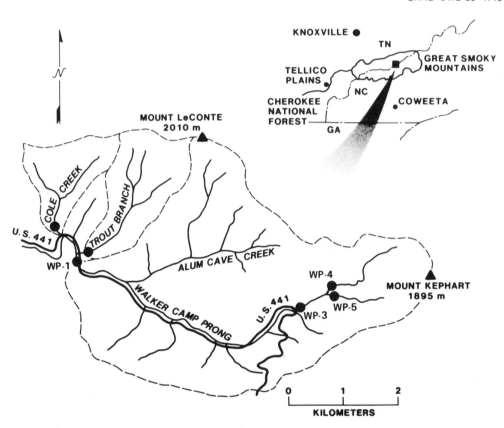

FIGURE 1.   Map of the study area showing the four sampling sites: Cole Creek, Trout Branch, Walker Camp Prong 4, and Walker Camp Prong 5.

Table 1
MEAN WATER QUALITY PARAMETERS

Site	pH	ANC (μeq · l^{-1})	SRP (μg · l^{-1})	TSP (μg · l^{-1})	TM-Al (mg · l^{-1})	OM-Al (mg · l^{-1})
CO	6.4	19.9	1.2	3.1	0.018	0.015
TB	5.0	−15.3	0.5	1.5	0.133	0.012
WP-5	5.7	−0.1	0.7	1.9	0.027	0.009
WP-4	4.5	−30.8	1.3	3.3	0.242	0.050
SC	6.5	44.4	2.0	3.7	0.005[a]	0.005[a]
HL	4.8	−8.6	1.1	3.6	0.327	0.014
WP-1	6.2	25.8	1.2	2.4	0.040	0.020
WP-3	4.7	−15.5	0.9	3.1	0.168	0.035

Note: Includes pH, acid-neutralizing capacity (ANC), soluble reactive P (SRP), total soluble P (TSP), total monomeric Al (TM-Al), and organic monomeric Al (OM-Al) measured at matched sites. CO — Cole Creek, TB — Trout Branch, SC — Sugar Cove, HL — Hemlock Creek, and WP — Walker Camp Prong.

[a]   At or below detection limit.

3) and Sugar Cove (SC)/Hemlock Creek (HL), were used only for comparisons of epilithic bacterial production. WP-1 and WP-3 are in the GSMNP and are downstream sites on Walker Camp Prong (Figure 1). WP-1 is farther downstream than WP-3 and is less acidic, with greater ANC and lower concentrations of Al (Table 1). SC and HL are second-order tributaries of the North River in the Cherokee National Forest (Figure 1), located about 60 km southwest of the GSMNP. SC is the less acidic of the two sites, with higher ANC and phosphorus concentrations and lower Al concentrations (Table 1). HL is acidified as a result of the exposure of pyrites to oxidation during road construction in its headwaters.

B. Experimental Design

1. Bacteria on Decomposing Leaves

Because of the uncertainty as to the length of time decomposing leaves have been in streams (and hence the degree of microbial colonization) and our desire to examine bacterial production on decomposing material over time, we placed mesh bags containing autumn-shed leaves at each stream site and sampled them at intervals over a 15-week period. Sugar maple (*Acer saccharum*) leaves were collected in autumn, leached in distilled water for 1 d, and air dried for a minimum of 7 d in a greenhouse. Nylon mesh bags (20 × 20 cm, 4-mm mesh) were filled with 5- (±0.05) g samples of leaves. Strings of ten bags were staked in five pools in each stream on August 27th (TB and CO) and August 29th (WP-4 and WP-5). From each of the 20 pools, 1 bag was collected at 7, 13, 28, 49, 70, 84, and 105 d after initiation of the experiment. No samples were taken after 105 d (15 weeks) due to the formation of ice in the streams and limited access to the sites and the bags. Litter bags were transported to the laboratory on ice in sealed plastic bags filled with stream water from the collection site.

In the laboratory, leaf material from each bag was gently rinsed in the stream water contained in the bag to remove adhering debris and macroinvertebrates. Disks (1 cm in diam) were cut from the leaf material and measured for bacterial production using tritiated thymidine incorporation into DNA; bacteria were counted using acridine orange direct counts.

Samples of leaf material were also taken for measurement of oxygen consumption, mass loss, ATP, P, N, and total Al content. Results of these measurements are reported elsewhere.[30]

2. Epilithic Bacterial Transplants

Rocks with their associated bacterial communities were incubated either in water from the site from which they were collected or in water from the matched site with differing pH. Thus, the effect of short-term changes in water chemistry could be compared to the effects of long-term exposure to acidic conditions. A total of four short-term experiments using rocks and water from different pairs of sites were run and analyzed as 2 × 2 factorial designs, with the origin of the rock as one factor and the origin of the water as the second. Tritiated thymidine incorporation into DNA was measured in short-term incubations (20 to 60 min) of rocks in jars containing stream water.

In a second series of experiments, effects of longer exposures to acidic conditions were examined. Rocks from CO were moved into TB, for periods of 23, 72, and 96 h. The rocks were then incubated with tritiated thymidine for 20 min and the results compared to those for controls moved from one site on CO to another site on the same creek.

C. Measurements of Thymidine Incorporation and Bacterial Numbers

Tritiated thymidine incorporation into DNA was measured as an indicator of bacterial production on the leaf material[11] and on rock surfaces. Five leaf disks cut from five different leaves from each bag were placed in a 50-ml round-bottom centrifuge tube containing 5 ml of water from the site at which the bag was taken. The tubes were equilibrated to stream temperature, and unlabeled (0.64 mol) and tritiated (0.16 nmol) thymidine (NET-250) (specific activity 2.886×10^{12} Bq · mmol^{-1}) was added to each tube. The disks were incubated for 20 min, after which 0.5 ml of 37% formaldehyde was added to the tubes. The liquid in the tubes was discarded and the leaf disks were rinsed three times with distilled water. The extraction of the labeled DNA from the leaf material followed the methods of Findlay et al.[10] The leaf disks were extracted overnight with 5 ml of 0.3 N NaOH, 1% sodium dodecyl sulfate (SDS), and 25 mM ethylene diamine tetraacetic acid (EDTA). The solution was neutralized and 5 ml of 10% trichloroacetic acid (TCA) and unlabeled DNA was added to the tubes. The tubes were centrifuged at $20,000 \times g$ for 15 min, and the leaf material and the precipitate were collected on Gelman 0.45-μm pore size filters. Next, 5 ml of 5% TCA was added to the filter and leaf material and extracted for 30 min at 100°C. Of the 5% TCA, 1-ml subsamples were analyzed for radioactivity in a liquid scintillation counter, and external standardization was used to correct to disintegrations per minute (dpm). Significant amounts of labeled thymidine appeared in the non-DNA fractions, perhaps due to the high concentration of thymidine added. The uptake reported is the incorporation into the DNA fraction only.

The uptake of tritiated thymidine by bacteria on rock surfaces was also determined using a modification of the technique of Findlay et al.[10] Small rocks (surface area <100 cm^2) were placed in wide-mouth polycarbonate jars, and sufficient stream water (50 to 70 ml) was added to totally immerse the rocks. Tritiated thymidine (final concentration 1.33×10^4 Bq · ml) was added to the jars and incubated within 2° of stream water temperature for 20 to 30 min. Incubation was terminated by addition of formaldehyde (0.02 ml of 37% formaldehyde · ml^{-1} stream water). The water in the jars was discarded and the rocks and jars were rinsed with distilled water three times. Between 50 and 70 ml of the NaOH-SDS-EDTA mixture was then added to the jars, and the rocks were extracted at room temperature overnight. Subsamples (20 ml) of the extractant were processed as described above for determination of the incorporation of thymidine into DNA. Isotope dilution experiments have not been completed for epilithic bacteria, thus the data are reported only as thymidine incorporation into DNA, not as bacterial production.

On each sampling day during the leaf decomposition experiment, additional leaf disks were taken from one of the sites for isotope dilution measurements. Thymidine incorporation was measured at additional concentrations of unlabeled thymidine (0 to 2.72 nmol added to 5 ml of incubation water) on these occasions. Data were transformed and analyzed to determine if the x-intercept was significantly different from zero. If the regression of the inverse of the dpm measured in DNA on the concentration of added thymidine was not significant, the regression was rerun after dropping the data from the highest concentration of added unlabeled thymidine. At high concentrations of added unlabeled thymidine the tritiated thymidine was significantly diluted. Thus, the actual dpm taken up were often very low and the blanks were similar to the samples. Thus, the error due to subtracting the blank was greatest at highest thymidine concentration.

A conversion factor derived from experimental and theoretical evidence[33] [3.0×10^{18} cells per mole of thymidine incorporated into DNA (2.4×10^{18} cells per mole of thymidine incorporated into ice-cold TCA divided by 0.8, the fraction of ice-cold TCA precipitate in the "purified" DNA fraction)] was used to relate thymidine incorporation to bacterial production because an empirical conversion factor cannot be reliably determined for bacteria in sediments or on leaf surfaces. After the conclusion of the experiment, we found that there was a sampling interval, weeks 2 to 4, over which we could make an estimate of the conversion factor. However, because we did not originally plan to estimate the conversion factor we did not sample in the optimal manner to do so. Also, there was probably grazing during the interval; thus, the derived conversion factor must be viewed as only a crude estimate. A review[34] of recent papers suggests that the theoretical factor we used is a conservative estimate of the factor and that actual values in a freshwater plankton system have been estimated using empirical methods[35] to be 4.7×10^{18} to 18.3×10^{18} cells per mole.[34] Production in terms of C was estimated using a recently derived conversion factor[36] of 0.35 pg C $\cdot$ um^{-3} and a cell volume of $0.222 \cdot \mu m^{-3}$, based on cell volume estimates made in a preliminary experiment.

DNA recovery using the NaOH-SDS-EDTA mixture on leaf material was estimated to be 76%, similar to recovery from sediments.[10] Thymidine incorporation data reported are uncorrected for the incomplete recovery, but calculation of bacterial production for bacteria on leaves is based on the corrected data. Specific growth rate (SGR) for the bacterial community on the leaves was calculated using the numbers of bacteria and the estimated cellular production rate. Doubling times were calculated by dividing 0.693 by SGR.

To determine if the factor we chose to convert thymidine incorporation into numbers of bacteria was reasonable, we compared estimates of bacterial production on leaf material derived from the thymidine data to estimates derived from the bacterial counts. The number of bacteria produced between sampling dates was calculated from the product of the mean of the thymidine-based production rates at the start and end of the interval and the length of the interval. This estimate was added to the number of bacteria present at the start of the interval to give a prediction of the number of bacteria present at the end and compared with changes in numbers of bacteria over that time. The initial number of bacterial cells and the production rate were assumed to be zero.

D. Water Chemistry

Samples for water chemistry were collected once a month at each site and at the time of each leaf bag collection. Samples were analyzed for ANC, SRP, TSP, total Al, and inorganic monomeric Al. Conductivity, pH, and temperature were also measured.

Samples were stored at 4°C until analyzed. Unfiltered samples were analyzed for ANC by titration with strong acid and Gran plot[37] analysis. Total Al[38] and inorganic monomeric Al[39] were measured by atomic absorption spectrophotometry using a graphite furnace.

Field pH was measured potentiometrically using an Orion (Model 221) portable pH meter with an Orion "Ross" glass electrode. Specific conductance and temperature were measured in the field with a portable, conductivity/temperature meter (Cole Parmer Model PCM 1). Samples for P determinations were filtered through a 0.4-μm Nuclepore filter. SRP was analyzed using the ascorbic acid method, and TSP was measured using the ascorbic acid method following persulfate digestion.[40]

III. RESULTS

A. Isotope Dilution Associated with Leaf Material

We obtained interpretable kinetic data in 9 of the 12 experiments but found no evidence for significant dilution of added thymidine by internal or external pools of thymidine (Table 2). The regression of the inverse of the dpm incorporated into DNA on the concentration of added thymidine was significant 7 of 12 times when all the data were included (e.g., Figure 2a) and on two additional occasions when data from the highest concentration were eliminated (e.g., Figure 2b). On the remaining three occasions, the high variability in the data (e.g., Figure 2c) prevented a determination of an isotope dilution curve even after the data from the highest concentration were removed (Table 2). The y-intercept was never significantly different from zero in any of the experiments. In this analysis, the internal and external pools of thymidine are estimated by the negative of the x-intercept. If the y-intercept is zero then the x-intercept is also zero and there is no evidence for dilution of the isotope.

We estimated as a "worst case" the degree of isotope dilution in the five experiments where the x-intercepts, while not statistically different than zero, were negative (Table 2) and thus a positive estimate could be made of the size of the thymidine pools. Because we

Table 2
MEASUREMENTS OF ISOTOPE DILUTION

Week	Site	N	r	Significance (%)	y-intercept	SE ($\times 10^{-5}$)	x-intercept (nm)
1	CO	9	0.876	99	−6.5	6.7	0.43
1	WP-4	8	0.839	99	−4.8	3.8	0.31
2	TB	9	0.778	95	3.3	3.5	−0.59
2	WP-5	9	0.905	99	1.9	3.3	−0.22
4	CO	9	0.536	NS	—	—	—
4	WP-4	8	0.746	95	−1.5	3.3	0.14
7	TB[a]	9	0.771	99	−0.29	3.3	0.02
10	CO[a]	8	0.749	95	1.0	0.8	−0.35
10	WP-4	8	0.003	NS	—	—	—
12	TB	16	0.710	99	1.8	1.7	−0.47
15	CO	9	0.743	95	0.61	0.34	−0.46
15	WP-4	9	0.457	NS	—	—	—

Note: Includes number of points included in the regression (N), the correlation coefficient (r), the significance level, the y-intercept (units are inverse dpm), the standard error of the y-intercept (SE), and the x-intercept. CO — Cole Creek, TB — Trout Branch, and WP — Walker Camp Prong.

[a] Highest concentration not included in the analysis.

measured thymidine incorporation at an added concentration of 0.80 nmol, if we included this worst-case estimate for the size of the internal and external pools of thymidine (0.42 nmol), the calculated incorporation rates would have been about 50% higher. The mean for all nine estimates, including the four negative estimates, was 0.13 nmol. Therefore, use of all the data yields an estimated dilution of only 16%. However, again none of the individual estimates of the dilution are significantly different from zero.

Thymidine incorporation was estimated as the incorporation at 0.80 nmol added thymidine per sample, the concentration at which we have five replicates. No correction was made for isotope dilution. Data for thymidine incorporation reported below are not corrected for the estimated 76% recovery efficiency for the DNA. However, estimates of production presented in the discussion are based on data corrected for DNA recovery.

B. Thymidine Incorporation by Bacteria on Leaves

The rate of thymidine incorporation into DNA by bacteria on leaves was generally highest at CO (Figure 3) and lowest at either TB (Figure 3) or WP-4 (Figure 4). The range in thymidine incorporation rates we observed was 0.50×10^{-13} to 6.53×10^{-13} mol thymidine $\cdot$ cm^{-2} leaf area $\cdot$ h^{-1}. The highest rate was observed at CO, 12 weeks after the leaves were placed in the stream, and the lowest rate of thymidine incorporation was at WP-4, 15 weeks after the leaves were placed in the streams.

Thymidine incorporation was consistently higher in the higher pH stream (CO or WP-5) when compared with the matched low pH stream (TB or WP-4). At six of the seven sampling times the bacterial production in CO was significantly higher than that at TB at the 95% level or greater (Table 3). Also, at five of the six sampling times the thymidine incorporation at WP-5 was significantly greater, at the 90% levels or above, than that at WP-4. Overall, the mean thymidine incorporation rate during the experiment was significantly ($p < 0.05$) greater (two-way analysis of variance and Duncan's multiple range test) at CO (mean = 3.72×10^{-13} mol $\cdot$ cm$^{-2} \cdot$ h^{-1}) than at TB (mean = 1.26×10^{-13} mol $\cdot$ cm$^{-2} \cdot$ h^{-1}) and was significantly greater at WP-5 (mean = 1.33×10^{-13} mol $\cdot$ cm$^{-2} \cdot$ h^{-1}) than at WP-4 (mean = 0.758×10^{-13} mol $\cdot$ cm$^{-2} \cdot$ h^{-1}). The critical range for four means was 0.537×10^{-13} mol $\cdot$ cm$^{-2} \cdot$ h^{-1}.

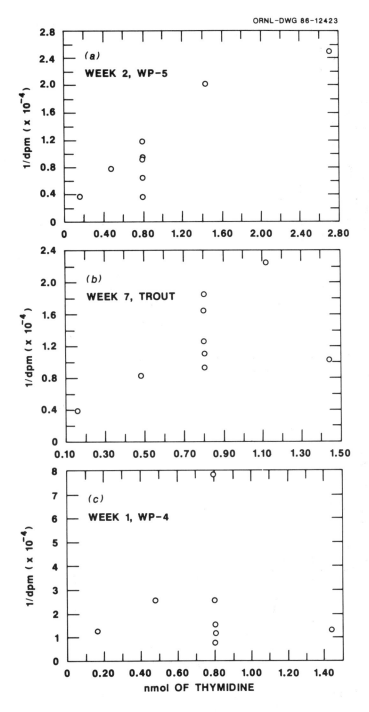

FIGURE 2. Isotope dilution plots from leaf material samples. (a) Data for week 2 at WP-5. (b) Data for week 7 at Trout Branch illustrating the deviation from the regression line of the data from the highest added concentration of thymidine. (c) Data for week 1 at WP-4 illustrating the large variability at 0.80 nmol added thymidine.

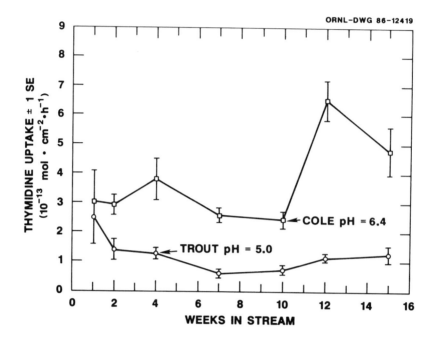

FIGURE 3.   Thymidine incorporation into DNA on decomposing leaf material at Cole Creek (pH 6.4) and Trout Branch (pH 5.0) in the fall.

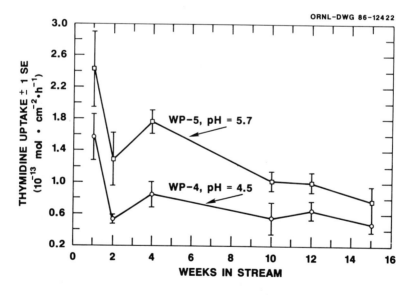

FIGURE 4.   Thymidine incorporation into DNA on decomposing leaf material at WP-4 (pH 4.5) and WP-5 (pH 5.7) in the fall.

The differences in rates of thymidine incorporation into DNA between matched higher and lower pH sites were evident 2 weeks after the leaves were placed in the streams (Table 3); however, at higher elevation sites the differences were not significant at 15 weeks, the final sampling time. With the exception of CO, thymidine incorporation was highest at the initial sampling, declined over the next 9 weeks, and stabilized after 10 weeks in the stream (Figures 3 and 4). At CO there was no decline in incorporation during the 1st 10 weeks (Figure 3).

Table 3
STUDENT *t* TESTS (t) FOR DIFFERENCES IN
THYMIDINE INCORPORATION BETWEEN COLE
CREEK (CO) AND TROUT BRANCH (TB) AND
BETWEEN WP-5 AND WP-4

Week	CO vs. TB			WP-5 vs. WP-4		
	Difference[a]	t	p > t	Difference[a]	t	p > t
1	0.53	0.42	NS[b]	0.86	1.74	NS
2	1.51	3.51	95%	0.75	2.45	90%
4	2.54	3.93	95%	0.93	4.86	95%
7	1.98	8.92	99.9%	ND[c]	ND	ND
10	1.71	6.17	99%	0.48	2.32	90%
12	5.43	8.92	99.9%	0.35	2.14	90%
15	3.59	4.66	99.9%	0.27	1.42	NS

[a] There were five replicates from each sampling time for each site. The units for the difference between the sites are 10^{-12} mol thymidine $\cdot$ cm^{-2} h^{-1}. Thymidine incorporation was never significantly higher in the low pH streams (TB or WP-4) than in the higher pH streams (CO or WP-5).
[b] NS = not significant.
[c] ND = not determined.

The thymidine incorporation data were used to estimate the total bacterial C production. Summing the production over the incubation period from week 1 to week 15, we estimate that the range in carbon production per litter bag was from 0.0277 g C at WP-4 to 0.155 g C at CO. Total C production per litter bag at TB and WP-5 were 0.0486 and 0.0543 g C, respectively.

C. Numbers of Bacteria on Leaf Material

The number of bacteria per gram of leaf material increased significantly during the 1st 4 weeks of the experiment (Figure 5), after which the numbers were fairly constant. The bacterial population was significantly higher ($p <0.5\%$) after 4 (t = 2.87), 10 (t = 2.48), and 15 (t = 2.40) weeks than at the initial sampling, 1 week after the leaves were placed in the stream. The range in numbers of bacteria ($\pm$ SE) was 5.57 ($\pm$ 0.62) $\times$ 10^{10} to 10.5 ($\pm$ 1.88) $\times$ 10^{10} bacteria per gram or 1.80×10^8 to $7.15 \times 10^8 \cdot$ cm^{-2}. The lowest standing stock was observed at the initial sampling and the highest was observed 10 weeks after the leaves were placed in the streams.

No significant differences in numbers of bacteria could be detected among the sites, probably due to the high variability, among samples at a site, of the bacterial counts (Figure 5). The mean ($\pm$ SE) number of bacteria on the leaf material over the entire course of the experiment only varied between 6.84 ($\pm$ 0.82) $\times$ 10^{10} and 8.67 ($\pm$ 1.60) $\times$ 10^{10} cells per gram. Even the nonsignificant trends in the data did not exhibit any pattern in numbers of bacteria in relation to pH. The highest mean number of bacteria was at one low pH site (TB) and the lowest mean number of bacterial cells was at the other low pH site (WP-4) (Figure 6).

D. Production of Bacterial Cells on Leaf Material

An indication of the accuracy of the factor used to convert thymidine incorporation into DNA to production of bacterial cells is given by a comparison of predicted and measured numbers of bacteria. The number of bacteria observed on the leaf material at weeks 2 and 4 was close to the number of bacteria predicted from the production rate and from the previous number of bacteria (Figure 7). Thus, in the initial stages of the decomposition, the

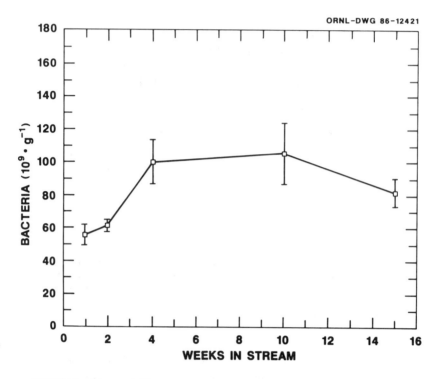

FIGURE 5.   Mean and SE of the number of bacteria observed on the leaf material over time. The data from all sites are averaged together.

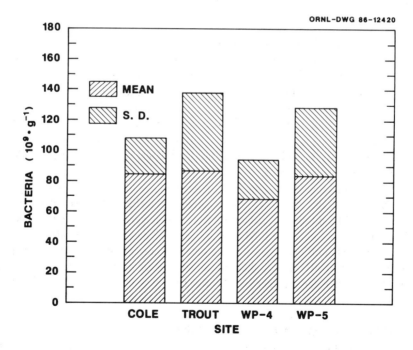

FIGURE 6.   Mean and SD of the number of bacteria observed on leaves at each site during the experiment. The data from all sampling times are averaged together.

conversion factor yields reasonable estimates of the production of bacterial cells. Predicted numbers of bacteria at week 1 were all much lower (between 58 and 79% lower) than the observed number. At three sites, (CO, TB, and WP-4), the predicted and observed numbers of bacteria at week 2 agreed to within 50% and at week 4 to within 21%. At the fourth site (WP-5), the predicted number of bacteria at week 2 was much greater than that observed, but, in comparison to the data from the other sampling times, the predicted number of bacteria appears to be a more reasonable estimate than the observed number (Figure 7). At week 4 prediction was much better, agreeing to within 21%.

After week 4 the measured numbers of bacteria on the leaf material began to decline so the predicted numbers consistently were two or more times higher than those measured (Figure 7). The mean over the four sites for the predicted number of bacteria was 70% lower at week 1, 29 and 6% higher at weeks 2 and 4, and 151 and 159% higher at weeks 10 and 15 than the measured values.

The SGR (Table 4), based on production rates and the bacterial counts, were faster at the higher pH sites than at the matched lower pH sites. The mean SGR at CO was greater than that at TB and the mean SGR at WP-5 was greater than that at WP-4. Specific growth rate varied between 0.0952 to 0.0114 d^{-1} for the individual sites and times. Mean SGR varied from 0.0723 to 0.0273 d^{-1} (Table 4). Doubling times (0.693/SGR), calculated from specific growth rates at individual sample times and sites, ranged from 7.3 d at CO after 15 weeks to 61 d at WP-4 after 10 weeks. Cell production rates followed the same pattern and were highest at CO and lowest at WP-4 (Table 4).

We used our data on numbers of bacteria and thymidine incorporation rates at weeks 2 and 4 at all four sites to estimate an empirical conversion factor for converting thymidine incorporation rate to production of bacterial cells. We used these data because the largest increase in bacterial numbers occurred between weeks 2 and 4. The estimate of the conversion factor was 5.31×10^{18} (SE of 1.43×10^{18}, n = 4) cells per mole thymidine. Due to the high variability of the bacterial counts, the limited amount of data on which the counts are based, and the probable loss of bacteria due to grazing and other factors during the interval, this is only a very crude estimation of the conversion factor.

E. Thymidine Incorporation by Epilithic Bacteria

Tritiated thymidine incorporation on rock surfaces ranged from 0.268×10^{-14} to 11.98×10^{-14} mol $\cdot$ cm$^{-2} \cdot$ h^{-1}. The highest rates of uptake were observed at CO and the lowest rates of uptake were observed at WP-4. In each of the short-term experiments the effect of the source of the rocks was significant ($p < 0.05$) and the effect of the source of the incubation water was not significant (Figure 8). A 3-h preexposure of the rocks and associated epilithic community to changes in water chemistry slightly increased thymidine incorporation into DNA in both controls and treatments subjected to reduced pH. Again, after the 3-h preexposure there was no significant difference between the controls and the treatments with reduced pH (Figure 8). Thus, it was evident that bacterial production was not affected by these short-term (0.5- to 3-h) changes in water chemistry. However, there were significant differences among the sites, with significantly higher rates of tritiated thymidine incorporation into DNA at the higher pH sites (Table 5, Figure 8).

The longer term transplants from CO (higher pH) into TB (low pH) yielded results similar to those of the short-term experiment for periods up to 24 h. In transplants of <24 h the transplanted rocks usually had lower uptake than the controls, but the differences were not significant (Table 6). For periods of >72 h, significant differences between rocks transplanted into the lower pH stream and the controls (rocks moved from one location to another in the higher pH stream) began to appear (Table 6).

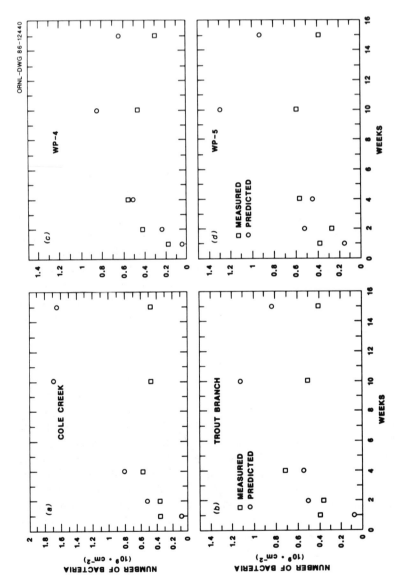

FIGURE 7.   Predicted and measured number of bacteria on the leaf material at each site during the experiment. (a) Cole Creek, (b) Trout Branch, (c) WP-4, and (d) WP-5.

Table 4
MEAN BACTERIAL PRODUCTION, SPECIFIC GROWTH RATES (SGR), BACTERIAL DOUBLING TIME (DT), AND THE RATE OF Al ACCUMULATION ON THE LEAF MATERIAL

Site	Production[a] (10^6 cells $\cdot$ cm^{-2} $\cdot$ d^{-1}) Mean	SE	SGR[a] (d^{-1}) Mean	SE	DT (d)	Al accumulation rate (mg Al $\cdot$ g^{-1} $\cdot$ d^{-1}) Slope	SE
CO	32.2	3.94	0.072	0.008	10.1	0.034	0.004
TB	13.5	2.73	0.031	0.008	29.4	0.113	0.008
WP-4	7.85	1.92	0.027	0.140	43.4	0.095	0.008
WP-5	13.8	2.79	0.034	0.008	25.8	0.067	0.005

[a] Mean for production is based on data from all samples. Mean SGR is based on data collected during weeks when both numbers of bacteria and bacterial production were measured (weeks 1, 2, 4, 10, and 15), doubling time is 0.693/SGR.

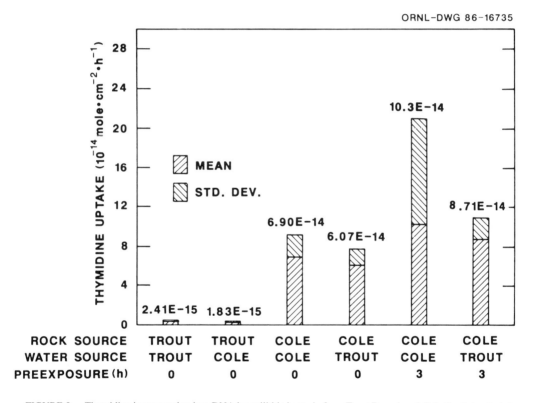

ORNL-DWG 86-16735

FIGURE 8. Thymidine incorporation into DNA by epilithic bacteria from Trout Branch and Cole Creek incubated in water from both sites.

IV. DISCUSSION

A. Bacteria on Leaf Material

The added thymidine was not significantly diluted by internal or external pools of thymidine during measurement of thymidine incorporation by bacteria on leaf material. Dilution of added thymidine during measurement of bacterial production was reported to be significant in some systems, primarily in sediments,[9,10] and not significant in others, primarily in planktonic systems.[33] In these experiments we used relatively high concentrations of thy-

Table 5

**MEAN THYMIDINE INCORPORATION INTO DNA BY
EPILITHIC BACTERIA ON ROCK SURFACES AT SITES
WITH DIFFERENT pH AT BASE FLOW**

Date	Station[a]	pH	Number of replicates	Thymidine incorporation (10^{-14} mol · cm^{-2} · h^{-1})
10/2/84[b]	WP-5	5.8	6	0.622
	WP-4	4.6	6	0.268
10/20/84[b]	SC-2	6.2	6	5.75
	HL-2	4.8	6	3.00
11/6/84	WP-1	6.6	16	1.10
	WP-3	4.9	8	0.547
4/1/85	CO	6.4	6	8.44
	TB	5.0	6	0.27

[a]   Stations indicated as WP are on Walker Camp Prong. Stations indicated as SC
are on Sugar Cove and stations indicated as HL are on Hemlock Creek. On all
sampling dates there was significantly greater ($p < 0.05$) thymidine incorporation
by epilithic bacteria at the higher pH site, but there was never a significant
difference associated with the source of the incubation water.

[b]   Data from Palumbo et al.[1]

Table 6

**LONG-TERM TRANSPLANT EXPERIMENTS FROM COLE CREEK
INTO TROUT BRANCH**

Date[a]	Hours	Treatment	Mean (10^{-14} mol · cm^{-2} · h^{-1})	N	Significance level
4/24/86	23	Transplant	7.67	6	NS[b]
		Control	11.98	6	
5/3/86	24	Transplant	2.67	10	NS
		Control	2.99	10	
5/13/86	72	Transplant	1.35	10	99%
		Control	4.08	10	

[a]   Date given is the day that the bacterial production was measured. Rocks were incubated in
the streams for the specified number of hours in open plastic trays (one per site).

[b]   NS = not significant.

midine, and no significant isotope dilution was detected. Thus, thymidine incorporation was
calculated on the basis of the specific activity of the thymidine added in the experiment.
Any error introduced by a small amount of isotope dilution is insignificant compared to the
possible error caused by the use of the theoretical conversion factor to convert from thymidine
incorporation to bacterial production. Theoretical conversion factors appear to underestimate
actual conversion factors determined empirically;[34] thus, the data presented here for bacterial
production and doubling times are conservative estimates.

Bacterial production on the leaf material was significantly lower in the more acidic streams
(Figures 3 and 4) and the reduced rate probably contributes to the slower decomposition we
observed under acidic conditions.[30] The rate of total Al accumulation (Table 4) by the leaf
material and associated bacteria was much faster at the low pH sites than at the higher pH
sites.[30] Coincidence of low bacterial production rates and high Al accumulation at low pH
may indicate an Al inhibition of bacterial metabolism at these highly acidic sites. Al is an
important factor in the toxicity of acidic waters to fish.[41,42]

There appear to be two pools of Al accumulating in association with decomposing leaf material. Total Al accumulates at a fairly constant rate over at least a 15-week period,[30] while oxalate extractable (amorphous) Al increases initially but stabilizes after about 4 weeks.[11] One or both of these pools may be associated with the reductions in bacterial production, perhaps due to adsorption to the bacterial cell surface and subsequent physiological interference.

The differences in the rate of bacterial production among the sites are clearly related to differences in the degree of acidification rather than to differences in P concentration. In an earlier study we found that thymidine incorporation into DNA by bacteria on leaf material in TB was significantly lower than that in CO.[11] However, because the P concentration was higher in CO, the effect could not unequivocally be attributed to acidification. However, in this study we not only found differences between CO and TB but also between WP-5 and WP-4. The bacterial production rate at WP-5, the higher pH site, was consistently higher than that at WP-4, despite the fact that WP-4 has higher SRP and TSP than WP-5 (Table 1).

The estimates of thymidine incorporation into DNA from this litter bag experiment are similar to those made in our earlier studies using material incubated in chambers. In the litter bag experiments our estimates of bacterial production, in terms of thymidine incorporation into DNA not corrected for the 76% recovery efficiency, ranged from 5.0×10^{-14} to 65.3×10^{-14} mol $\cdot$ cm$^{-2} \cdot$ h^{-1} compared to 20.5×10^{-14} to 61.1×10^{-14} mol $\cdot$ cm^{-2} $\cdot$ h^{-1} in the earlier chamber experiments.[11] The litter bag experiments were run in the fall and winter, and water temperatures fell from 15 to 6°C over the course of the study. The chamber experiments were run in the summer and water temperatures ranged from 12 to 16°C. Thus, the wider range of bacterial production rates could be a result of the wider range of temperatures during the litter bag experiments. The similarity of rates occurs despite the fact that in the chamber experiments tritiated thymidine was used at a specific activity of approximately 2.9×10^{12} Bq $\cdot$ mmol^{-1}, and in the litter bag experiment the thymidine was diluted with unlabeled thymidine to a specific activity of approximately 5.55×10^{11} Bq $\cdot$ mmol^{-1}.

Our estimated mean doubling times are long (Table 4) and our estimated specific growth rates are low (Table 4) compared to rates estimated for planktonic bacteria in lakes, estuaries, and oceans.[33,43-46] However, our doubling times (10 to 43 d) are comparable to estimates (12 to 32 d) made for bacteria on decomposing leaf material in an artificial stream system[47] and for estimates (20 d) made for bacteria in sediments.[9]

Our estimates of bacterial production, doubling times, and specific growth rate are probably conservative due to the use of a conservative theoretical factor[34] of 3.0×10^{18} for converting thymidine incorporation to bacterial production. Our crude empirical estimate of the conversion factor (5.31×10^{18}) was higher by a factor of 1.76 than the factor used in the calculations of doubling times and specific activity. Therefore, use of this empirical conversion factor would change the estimates of doubling times, etc. by a similar factor, yielding estimates of 5.7 to 24.8 d for the doubling time. These estimates are still extremely long compared to those found in planktonic systems.

The calculation of a specific growth rate implies that all the bacteria in a sample are growing at the same rate. The growing population of plankton bacteria may consist of only 10 to 50% of the total population in oceanic samples,[48] and a large proportion of the bacteria attached to the leaf material may not have been growing. Thus, the growth rates for the active proportion of the population could be much higher than the specific growth rate presented above. The difference between planktonic systems and bacteria growing on the leaf material may be due to differences in the proportion of bacteria that are actively growing.

A relatively small amount of C was incorporated into bacterial biomass over the course of the 15-week experiment. The highest total bacterial C production was at CO, where the production per litter bag was 0.155 g compared to a starting mass of 2.2 g of leaf C. The amount of C respired was greater than the amount of leaf C converted to bacterial biomass. Using the estimates of bacterial C production given above and estimates of total microbial respiration on the leaf material,[30] we calculated a range in production-to-respiration ratio from 0.40 at CO to 0.10 at WP-4. However, the measurement of respiration included respiration by bacteria and eukaryotic organisms (fungi, protozoa) associated with the leaf material. Thus, although this calculation does illustrate that a significant proportion of the detrital C processed by the bacteria is being respired, it represents a lower limit of the conversion efficiency for the transformation of detritus into bacterial C.

No significant differences exist among the sites in the number of bacteria on leaf material, and these values are similar to those reported in other studies. Carpenter et al.[2] also found no significant differences among sites related to pH or other variables despite the fact that decomposition rates were slower at their stream sites affected by acid mine drainage. Carpenter et al.[2] found numbers of bacteria in the range of 5.8×10^9 to 8.0 to 10^{10} cells per gram (oven-dried weight) and that numbers increased rapidly, up to tenfold, over the 1st week and only slightly thereafter. Our first sampling was at 1 week and this may be the reason we did not see the low initial numbers. However, our subsequent observations of 6.84×10^{10} to 8.67×10^{10} cells per gram agree well with those of Carpenter et al.[2] for leaf material. Our observations are also close to those for decomposing *Spartina*[49] (up to 2.98×10^{10} cells per gram) and *Thalassia*[50] (from 0.9×10^{10} to 6.0×10^{10} cells per gram) as well as being close to data for other decomposing plant material in estuarine and marine environments.[51] Our bacterial densities, 1.80×10^8 to $7.15 \times 10^8 \cdot cm^{-2}$, are higher than those reported for oak leaves in an artificial stream system[47] (1.4×10^7 to 2.37×10^7 cells per square centimeter).

The calculated predictions of numbers of bacteria based on production rate indicate that during the later stages of the study there likely was a substantial loss of bacteria from the leaf material. Predictions for numbers of bacteria at week 1 were much lower than the actual numbers observed, probably due to the implicit assumption in our prediction method that the initial population and the production rate were zero. However, predictions for weeks 2 and 4 agreed with measurements. Predictions of numbers of bacteria for weeks 10 and 15 are substantially higher than observed numbers. The bacteria are probably lost to grazing, sloughing, or cell lysis, or they are released from leaf material into the water.

The potential for grazing by macroinvertebrates was greatest during the time of the highest apparent bacterial loss rates. The numbers of macroinvertebrates associated with the leaf material increased over the 1st 4 weeks of the experiment and were relatively stable thereafter,[30] a pattern very similar to that observed for the bacteria (Figure 5). Thus, during the last 11 weeks of the experiment, both the apparent loss of bacteria and the number of macroinvertebrates were greatest. The importance of protozoa or other microfauna was not assessed, but grazing by these organisms may be significant.

Although the bacterial production was clearly greater at the higher pH sites, differences in leaf weight loss among the sites were not as pronounced.[30] Therefore, reduced bacterial decomposition at lower pH sites may have been compensated for by an increase in the weight loss attributable to fungi or macroinvertebrates. There have been reports that acidification favors fungal decomposition[23,52] and that cellulase from stream fungi may function best under acidic conditions.[53] However, it is unlikely that fungi compensate fully for the reduction in bacterial production because decomposition and microbial respiration under acidic conditions in these[30] and other streams[15,54] do appear to be significantly reduced. Even lignocellulose, which is thought to be decomposed by fungi in terrestrial systems,[55-57] appears to be decomposed primarily by bacteria in aquatic systems.[58] Thus, it appears that acidification reduces bacterial production on leaf material and decomposition of bulk leaf material and specific components of leaves.

B. Epilithic Bacteria

Thymidine incorporation into DNA by the epilithic bacteria was significantly affected by acidification. The rate of thymidine incorporation by the epilithic bacteria in streams with pH ≥ 5.7 was significantly greater than that in matched streams with pH ≤ 5.0. Analysis of variance of the four short-term (≤ 3-h) transplant experiments indicated that in each experiment the source of the epilithic community significantly affected the rate of thymidine incorporation, with approximately 100% higher uptake rates by the bacteria from the higher pH site than by the bacteria from the matched lower pH site (Table 5, Figure 8). However, in all four experiments, the source of the sample water had no effect on the uptake rates. Thus, exposing epilithic bacteria to water with pH ≤ 5.0 and Al concentrations ≥ 0.10 mg $\cdot$ l^{-1} for ≤ 3 h did not significantly reduce the thymidine incorporation rate. Similarly, exposing epilithic bacteria from a low pH site to water from a higher site with low Al concentration (≤ 0.015 mg $\cdot$ l^{-1}) did not significantly increase the rate of thymidine incorporation. Evidently, the negative effects of acidification on the bacterial community involve relatively long-term stress and are not quickly reduced with more favorable conditions.

Exposures of >48 h to reduced pH and increased Al levels were required before there were significant reductions in epilithic bacterial production (Table 6). In experiments with exposures to acidified conditions of 23 to 24 h, there was no significant reduction in the rate of uptake of thymidine incorporation into DNA. The effects of these changes on pH and Al are much slower in developing than are effects of trace metals on bacteria. For example, Cu can significantly reduce the uptake of amino acids by bacterioplankton communities in exposures of less than 5 h.[59] Our results are comparable to those of Osgood[60] who found that exposures of >48 h to acidic water were necessary to significantly reduce thymidine incorporation by epilithic bacteria in Adirondack streams. Because the length of time required to cause a significant reduction in bacterial production was so long, both direct toxicity and indirect effects could potentially reduce bacterial production at the low pH site. One possible indirect effect is reduction in grazing pressure at the low pH sites.[32]

The epilithic prokaryotic (bacteria, blue-green algae) community appears to be more sensitive to acidification than the epilithic eukaryotic autotrophic community. We observed reductions in epilithic bacterial production at stream sites with pH ≤ 5.0, but there is no observed decrease in primary production[32] at many of the same low pH sites. Also, chlorophyll *a* concentrations were higher at some of the low pH sites[32] that were found to have lower total ATP concentrations,[1] indicating a reduction in total microbial biomass at low pH despite an increase in the eukaryotic autotroph biomass. Even among the autotrophic periphyton, the prokaryotic cyanobacteria are most affected by acidic conditions.[32] In other studies, there was less organic C and fewer microbial cells found on stones in an acidic stream in England than in a less acidic stream.[27] However, again indicating the relatively greater sensitivity of the bacterial community compared to the algal community, there were higher concentrations of plant pigments at the more acidic site than at a less acidic site.

The rate of thymidine incorporation per unit surface area by the epilithic bacteria was generally lower than the rate of uptake by the bacteria on the leaf surfaces. Epilithic bacterial incorporation rates were in the range of 0.14×10^{-13} to 1.20×10^{-13} mol $\cdot$ cm^{-2} $\cdot$ h^{-1} and incorporation rates by the bacteria on the leaf surfaces were in the range of 0.4×10^{-13} to 6.5×10^{-13} mol $\cdot$ cm^{-2} $-$ h^{-1}. This is not surprising because bacteria on leaf surfaces have access to additional sources of C and nutrients from the leaf material.

V. CONCLUSIONS

The bacteria growing on leaf material and on rock surfaces in these mountain streams appear to be significantly affected by acidification. Incorporation of tritiated thymidine by the bacteria on leaf and rock surfaces is significantly lower at low pH sites (≤ 5.0) than at matched higher pH sites (≥ 5.5). The effect of acidification is not immediate; results from transplant experiments indicated that it takes from 72 h to greater than 1 week for significant effects of changes in water chemistry to have an effect on bacterial production. Because the time required to generate significant reductions in bacterial production with exposure to acidic stream water (pH ≤ 5.0) is relatively long, indirect effects (e.g., reductions in grazing activity) may play a role in reducing bacterial production. However, even in the absence of macroinvertebrate grazers, there are significantly lower rates of thymidine incorporation by bacteria on leaf material in acidic streams. Al is implicated in the acidification effects on the microbial community; the rate of Al accumulation on leaves is significantly greater at the low pH sites than at the matched higher pH sites. On an areal basis (per square centimeter), rates of thymidine incorporation into DNA for bacteria growing on leaf surfaces are greater than rates for epilithic bacteria. The doubling times of bacteria growing on the decomposing leaf material are extremely long compared to those in planktonic systems when bacterial production is calculated using either a theoretical literature value for the conversion of thymidine incorporation to production of bacterial cells (3.0×10^{18} cells per mole) or a crude empirical estimate (5.31×10^{18} cells per mole) derived from our experiments. Additional experimentation is ongoing to examine the effects of concomitant exposures to low pH and high Al concentrations vs. exposure only to low pH and the effects of reductions in grazing on epilithic bacterial production.

ACKNOWLEDGMENTS

We thank M. A. Bogle and R. R. Turner for help with the chemistry data, L. A. Ferren and D. M. Genung for technical assistance, and H. L. Boston and R. B. Cook for reviews of the manuscript. We also thank the U.S. Park Service for their cooperation in our sampling program in the Great Smoky Mountains National Park. This research was supported by the Electric Power Research Institute under contract No. RP2326-1 with the Oak Ridge National Laboratory operated by Martin Marietta Energy Systems, Inc., under contract No. DE-AC05-84OR1400 with the U.S. Department of Energy. Publication No. 3164, Environmental Sciences Division, Oak Ridge National Laboratory.

REFERENCES

1. **Palumbo, A. V., Bogle, M. A., Turner, R. R., Elwood, J. W., and Mulholland, P. J.,** Bacterial communities in acidic and circumneutral streams, *Appl. Environ. Microbiol.,* 53, 337, 1987.
2. **Carpenter, J., Odum, W. E., and Mills, A.,** Leaf litter decomposition in a reservoir affected by acid mine drainage, *Okios,* 41, 165, 1983.
3. **Rosset, J. and Barlocher, F.,** Processing of leaf and needle litter in four streams of different alkalinity, *Verh. Int. Ver. Limnol.,* 21, 871, 1981.
4. **Suberkropp, K. and Klug, M. J.,** Fungi and bacteria associated with leaves during processing in a woodland stream, *Ecology,* 57, 707, 1976.
5. **Buttimore, C. A., Flannagan, P. W., Cowan, C. A., and Oswood, M. W.,** Microbial activity during leaf decomposition in an Alaskan subarctic stream, *Holarct. Ecol.,* 7, 104, 1984.
6. **Mulholland, P. J., Elwood, J. W., Newbold, J. D., and Ferren, L. A.,** Effect of leaf-shredding invertebrates on phosphorus spiraling in heterotrophic streams, *Oecologia,* 66, 199, 1985.
7. **Kaushik, N. K. and Hynes, H. B. N.,** The fate of the dead leaves that fall into streams, *Arch. Hydrobiol.,* 68, 465, 1971.
8. **Fuhrman, J. and Azam, F.,** Bacterioplankton secondary production estimates for coastal waters of British Columbia, Antarctica, and California, *Appl. Environ. Microbiol.,* 39, 1085, 1980.
9. **Moriarty, D. and Pollard, P.,** Diel variation of bacterial productivity in seagrass (*Zostera caprocorni*) beds measured by the rate of thymidine incorporation into DNA, *Mar. Biol.,* 72, 165, 1982.

10. **Findlay, S. E. G., Meyer, J. L., and Edwards, R. T.,** Measuring bacterial production via rate of incorporation of [^{3}H] thymidine into DNA, *J. Microbiol. Methods,* 2, 57, 1984.
11. **Palumbo, A. V., Mulholland, P. J., and Elwood, J. W.,** Microbial communities on leaf material protected from macroinvertebrate grazing in acidic and circumneutral streams, *Can. J. Fish. Aquat. Sci.,* 44, 1064, 1987.
12. **Francis, A. J., Quinby, H. L., and Hendrey, G. R.,** Effect of lake pH on microbial decomposition of allochthonous litter, in *Early Biotic Responses to Advancing Lake Acidification,* Hendrey, G. R., Ed., Butterworth, S., Boston, 1984, chap. 1.
13. **Andersson, G.,** Decomposition of alder leaves in acid lake waters, *Ecol. Bull.,* 37, 293, 1985.
14. **Friberg, F., Otto, C., and Svensson, B. S.,** Effects of acidification in the dynamics of allochthonous leaf material and benthic invertebrate communities in running waters, in *Ecological Impact of Acid Precipitation,* Drablos, D. and Tollan, A., Eds., SNSF Project, Oslo, Norway, 1980, 304.
15. **Hildrew, A. G., Townsend, C. R., Francis, J., and Finch, K.,** Cellulolytic decomposition in streams of contrasting pH and its relationship with invertebrate community structure, *Freshwater Biol.,* 14, 323, 1984.
16. **Petersen, R. C. and Cummings, K. W.,** Leaf processing in a woodland stream, *Freshwater Biol.,* 4, 343, 1974.
17. **Iversen, T. M.,** Disappearance of autumn shed leaves placed in bags in small streams, *Verh. Int. Ver. Limnol.,* 19, 1687, 1975.
18. **Hart, S. D. and Howmiller, R. P.,** Studies on the decomposition of allochthonous detritus in two southern California streams, *Verh. Int. Ver. Limnol.,* 19, 1665, 1975.
19. **Benfield, E. F. and Webster, J. R.,** Shredder abundance and leaf breakdown in an Appalachian mountain stream, *Freshwater Biol.,* 15, 113, 1985.
20. **Burton, T. M., Stanford, R. M., and Allan, J. W.,** Acidification effects on stream biota and organic matter processing, *Can. J. Fish. Aquat. Sci.,* 42, 669, 1985.
21. **Hendry, G. R., Baalsrud, K., Traaen, T. S., Laake, M., and Raddum, G.,** Acid precipitation: some hydrobiological changes, *Ambio,* 5, 224, 1976.
22. **Minshall, G. W. and Minshall, J. N.,** Further evidence on the role of chemical factors in determining the distribution of benthic invertebrates in the River Duddon, *Arch. Hydrobiol.,* 83, 324, 1978.
23. **Traaen, T. S.,** Effects of acidity on decomposition of organic matter in aquatic environments, in *Ecological Impact of Acid Precipitation,* Drablos, D. and Tollan, A., Eds., SNSF Project, Oslo, Norway, 1980, 340.
24. **Rao, S. S., Paolini, D., and Leppard, G. G.,** Effects of low-pH stress on the morphology and activity of bacteria from lakes receiving acid precipitation, *Hydrobiologia,* 114, 115, 1984.
25. **Geesey, G. G., Mutch, R., Costerton, J. W., and Green, R. B.,** Sessile bacteria: an important component of the microbial population in small mountain streams, *Limnol. Oceanogr.,* 23, 1214, 1978.
26. **Ladd, T. I., Costerton, J. W., and Geesey, G. G.,** Determination of the heterotrophic activity of epilithic microbial populations, in *Native Aquatic Bacteria: Enumeration, Activity and Ecology,* Costerton, J. W. and Colwell, R. R., Eds., American Society for Testing and Materials, Philadelphia, 1979, 180.
27. **Winterbourn, M. J., Hildrew, A. G., and Box, A.,** Structure and grazing of stone surface organic layers in some acid streams of southern England, *Freshwater Biol.,* 15, 363, 1985.
28. **Haack, T. K. and McFeters, G. A.,** Nutritional relationships among micro-organisms in an epilithic biofilm community, *Microb. Ecol.,* 8, 115, 1982.
29. **Lock, M. A. and Ford, T. E.,** Microcalorimetric approach to determine relationships between energy supply and metabolism in river epilithon, *Appl. Environ. Microbiol.,* 49, 408, 1985.
30. **Mulholland, P. J., Palumbo, A. V., Elwood, J. W., and Rosemond, A. D.,** Effect of acidification on leaf decomposition in streams, *J. North Am. Benthol. Soc.,* 6, 147, 1987.
31. **Huckabee, J. W., Goodyear, C. P., and Jones, R. D.,** Acid rock in the Great Smokies: unanticipated impact on aquatic biota of road construction in regions of sulfide mineralization, *Trans. Am. Fish. Soc.,* 104, 677, 1975.
32. **Mulholland, P. J., Elwood, J. W., Palumbo, A. V., and Stevenson, R. J.,** Effect of stream acidification on periphyton composition, chlorophyll, and productivity, *Can. J. Fish. Aquat. Sci.,* 43, 1846, 1986.
33. **Fuhrman, J. and Azam, F.,** Thymidine incorporation as a measure of heterotrophic bacterioplankton production in marine surface waters: evaluation and field results, *Mar. Biol.,* 66, 109, 1982.
34. **Scavia, D. G., Laird, A., and Fahnenstiel, G. L.,** Production of planktonic bacteria in Lake Michigan, *Limnol. Oceanogr.,* 31, 612, 1986.
35. **Kirchman, D., Ducklow, H., and Mitchell, R.,** Estimates of bacterial growth from changes in uptake rates and biomass, *Appl. Environ. Microbiol.,* 44, 1296, 1982.
36. **Bjornsen, P. K.,** Automatic determination of bacterioplankton biomass by image analysis, *Appl. Environ. Microbiol.,* 51, 1199, 1986.
37. **Gran, G.,** Determination of the equivalence point in potentiometric titrations, *Int. Congr. Anal. Chem.,* 77, 661, 1952.

38. **Barnes, R. B.,** The determination of specific forms of aluminum in natural water, *Chem. Geol.,* 15, 177, 1976.

39. **Driscoll, C. T.,** A procedure for the fractionation of aqueous aluminum in dilute acidic waters, *Int. J. Environ. Anal. Chem.,* 6, 267, 1984.

40. American Public Health Association, American Water Works Association, and Water Polution Control Federation, *Standard Methods for Examination of Waste and Waste Water,* 16th ed., American Public Health Association, Washington, D.C., 1985, 1268.

41. **Driscoll, C. T., Baker, J. P., Bisogni, J. J., and Schofield, C. L.,** Effects of aluminum speciation on fish in dilute acidified waters, *Nature,* 284, 161, 1980.

42. **Baker, J. P. and Schofield, C. L.,** Aluminum toxicity to fish in acidified waters, *Water Air Soil Pollut.,* 18, 28, 1982.

43. **Ducklow, H. W.,** Chesapeake Bay nutrient and plankton dynamics. I. Bacterial biomass and production during spring tidal destratification in the York River, Virginia, *Limnol. Oceanogr.,* 27, 651, 1984.

44. **Ducklow, H. W., Kirchman, D. L., and Rowe, G. T.,** Production and vertical flux of attached bacteria in the Hudson River plume of the New York Bight as studied with floating sediment traps, *Appl. Environ. Microbiol.,* 43, 769, 1982.

45. **Bell, R. T., Ahlgren, G. M., and Ahlgren, I.,** Estimating bacterio-plankton production by measuring ^{3}H-thymidine incorporation in a eutrophic Swedish lake, *Appl. Environ. Microbiol.,* 45, 1709, 1983.

46. **Ducklow, H. W. and Hill, S. M.,** The growth of heterotrophic bacteria in the surface waters of warm core rings, *Limnol. Oceanogr.,* 30, 239, 1985.

47. **Perkins, R. E., Elwood, J. W., and Sayler, G.,** *Detrital Microbial Community Development and Phosphorus Dynamics in a Stream Ecosystem,* ORNL/TM-9920, Oak Ridge National Laboratory, Oak Ridge, TN, 1986.

48. **Ducklow, H. W. and Hill, S. M.,** Tritiated thymidine incorporation and the growth of heterotrophic bacteria in warm core rings, *Limnol. Oceanogr.,* 30, 260, 1985.

49. **Rublee, P. A.,** Bacteria in a North Carolina Salt Marsh: Standing Crop and Importance in Decomposition of *Spartina alternifloria,* Ph.D. thesis, North Carolina State University, Raleigh, 1978.

50. **Rublee, P. A. and Roman, M. A.,** Decomposition of turtlegrass (*Thalassia testudinum* Konig) in flowing seawater tanks and litterbags: compositional changes and comparison with natural particulate matter, *J. Exp. Mar. Biol. Ecol.,* 58, 47, 1982.

51. **Fenchel, T. and Harrison, P.,** The significance of bacterial grazing and mineral cycling for the decomposition of particulate detritus, in *The Role of Terrestrial and Aquatic Organisms in Decomposition Processes,* Anderson, J. M. and Macfadyen, A., Eds., Blackwell, Oxford, 1976, 285.

52. **Hall, R. J., Likens, G. E., Fiance, S. B., and Hendrey, G. R.,** Experimental acidification of a stream in the Hubbard Brook Experimental Forest, New Hampshire, *Ecology,* 61, 976, 1980.

53. **Sinsabaugh, R. L., III, Benfield, E. F., and Linkins, A. E., III,** Cellulase activity associated with the decomposition of leaf litter in a woodland stream, *Okios,* 36, 184, 1981.

54. **Benner, R., Moran, M. A., and Hodson, R. A.,** Effects of pH and plant source on lignocellulose biodegradation rates in two wetland ecosystems, the Okefenokee Swamp and a Georgia salt marsh, *Limnol. Oceanogr.,* 30, 489, 1985.

55. **Kirk, T. K., Connors, W. J., and Ziekus, J. G.,** Requirements for a growth substrate during lignin decomposition by two wood rotting fungi, *Appl. Environ. Microbiol.,* 32, 192, 1975.

56. **Pugh, G. J.,** Terrestrial fungi, in *Biology of Plant Litter Decomposition,* Vol. 2, Dickinson, C. H. and Pugh, G. J., Eds., Academic Press, New York, 1974, 303.

57. **Whitkamp, M. and Ausmus, B. S.,** Processes in decomposition and nutrient transfer in forest systems, in *The Role of Terrestrial and Aquatic Organisms in Decomposition Processes,* Anderson, J. M. and Macfadyen, A., Eds., Blackwell, Oxford, 1976, 375.

58. **Benner, R., Moran, M. A., and Hodson, R. A.,** Biogeochemical cycling of lignocellulosic carbon in marine and freshwater ecosystems: relative contribution of procaryotes and eucaryotes, *Limnol. Oceanogr.,* 31, 89, 1986.

59. **Sunda, W. G. and Ferguson, R. L.,** Utilization of amino acids by planktonic marine bacteria: importance of clean technique and low substrate additions, *Limnol. Oceanogr.,* 29, 258, 1984.

60. **Osgood, M. S.,** Bacterial Communities in Acidic Adirondack Streams, Thesis, Rensselaer Polytechnic Institute, Troy, NY.

Chapter 7

ALGAL ASSEMBLAGES IN ACID-STRESSED LAKES WITH PARTICULAR EMPHASIS ON DIATOMS AND CHRYSOPHYTES

Sushil S. Dixit and John P. Smol

TABLE OF CONTENTS

I. INTRODUCTION

In eastern North America and Europe, precipitation has become increasingly acidic because of elevated emissions of sulfuric and nitric oxides from metal smelting, fossil fuel combustion, and other industrial processes. Long-range transport of these pollutants has been enhanced by the installation of tall stacks and the removal of particulates from emissions. The average precipitation pH (4.0 to 4.4) in these sensitive areas is about 100 times more acidic than the pH of natural rainwater.[1] This acidic precipitation has threatened a large number of lakes in acid-sensitive areas, and the resulting acidification of inland waters has become one of the most serious environmental problems of the 1970s and 1980s.

Acid-stressed lakes have the following characteristics: (1) they receive precipitation of mean annual pH of 4.5 or less,[2] (2) their drainage basins are underlain by siliceous bedrock,[3] and (3) the soils in their catchments are noncalcareous, and therefore have little buffering capacity.[4] Soil characteristics can override all other factors.[5] Topography also has an important influence on acidification, with headwater lakes located in igneous and metamorphic rocks being especially susceptible.[6]

Many acidification-related changes have been identified in algal communities, including shifts in species composition, and changes in species richness, diversity, and biomass. Because lake acidification is closely associated with increases in water transparency, high concentrations of toxic metals, changes in phosphorus concentrations, and shifts in zooplankton species composition, it is difficult to relate algal changes simply to depressions in lakewater pH.

Studies dealing with the effects of acidification on algal communities are primarily from field observations where little control of environmental variables is usually possible. Field studies are further complicated by the temporal and spatial succession in algal communities, and so integrated collections over relatively long time frames are required for holistic overviews. Moreover, field data require careful interpretation because most of these observations are based on correlations. It is therefore problematic to study the impacts of specific factors on algal communities without complementary laboratory studies that address questions dealing with physiological mechanisms. Unfortunately, such studies are often lacking.

Before any phycological review can be undertaken, some reference must be made on the taxonomic system that will be followed — there are several. Table 1 lists the class system that we will use. It is only a partial classification, which is restricted to the taxa that we will be discussing. Several textbooks are available which discuss algal systematics in detail.[7,8]

II. PHYTOPLANKTON ALGAL COMMUNITIES OF ACIDIC LAKES

Any attempt at providing a summary of phytoplankton changes associated with acidification is fraught with difficulties, as data are still lacking for any comprehensive overview. Assessment of the published literature is complicated further by differences in sampling techniques,[9] the taxonomic competence of observers,[10] and the fact that most phytoplankton exhibit marked seasonality,[11] and therefore short-term sampling intervals may be misleading. Nevertheless, some generalizations can be made.

A. Species Composition

Generally, in nonacidic oligotrophic lakes, the predominant planktonic algae are chrysophytes, greens, and diatoms, whereas blue-greens are rare.[12,13] As a lake acidifies, these algae are gradually replaced by dinoflagellates, and, to some extent, by cryptophytes,[14,15] but exceptions often occur. Experimentally, it has been shown that dinoflagellates and cryptophytes are more tolerant of rapid declines in pH than other classes.[15] The maximum changes in phytoplankton communities usually occur over a pH range of 4.7 to 5.6,[16] the same range where the bicarbonate buffering system is lost during lake titration.

Table 1
ALGAL CLASSIFICATION SYSTEM THAT IS FOLLOWED IN THIS CHAPTER

Class	Common name	Representative Genera
Cyanophyceae or Cyanobacteria	Blue-green algae	*Oscillatoria, Phormidium,* and *Merismopedia*
Chlorophyceae	Green algae	*Chlorella, Mougeotia,* and *Zygnema*
Dinophyceae	Dinoflagellates	*Peridinium, Gymnodinium*
Cryptophyceae	Cryptophytes	*Cryptomonas*
Bacillariophyceae	Diatoms	*Tabellaria, Frustulia, Eunotia, Navicula,* and *Pinnularia*
Chrysophyceae	Chrysophytes	*Dinobryon, Mallomonas,* and *Synura*

Note: This is an incomplete listing, only including the groups discussed in this chapter.

Various factors may be responsible for shifts in algal communities in acidified lakes, including physiological optima for low pH waters,[17,18] shifts in invertebrate grazing pressure,[19,20] as well as other changes in competitive coefficients. As will be discussed below, lake acidification can influence these factors directly or indirectly.

Peridinium inconspicuum is often a dominant dinoflagellate in clearwater acidic lakes,[14,16,21] and its abundance is usually simply attributed to its higher tolerance for acidic waters. The other common taxa in acidic lakes are *Gymnodinium uberrimum* and *Chlamydomonas sp.*[14,22] In acidic (pH = 4.9) Silver Lake, Adirondack Mountains, *Chlamydomonas* spp. were also the dominant taxa.[23] Meanwhile, in slightly acidic Ontario lakes (pH 5.5 to 6.2) *Chrysochromulina breviturrita* (Prymnesiophyceae) is a common plankter.[24]

Experimental acidification of Lake 223 in the experimental lakes area (Ontario) decreased the pH from 6.8 to 5.0 over an 8-year period, and dramatic changes were recorded in the phytoplankton.[25] Although chrysophytes declined slightly when pH dropped to about 6.1, they still dominated the plankton. At pH 5.9, various green, blue-green, and Peridineae dinoflagellate species increased, with compensatory declines in diatoms and chrysophytes. The development of a *Chlorella* bloom below the thermocline was primarily responsible for the increase in Chlorophyceae, and further shows how only sampling epilimnetic waters can be misleading. Once the pH had been lowered to about 5.6, a dramatic increase was observed in the acidic diatom *Asterionella ralfsii*. Subsequent declines in pH to 5.1 resulted in a dramatic decline in chrysophytes and their replacement by dinoflagellates (*Gymnodinium* and *Peridinium*), blue-greens (*Merismopedia* and *Chroococcus*), and to some extent by greens (*Chlorella* spp.).[25,26] Undoubtedly, many variables influenced the algal community in this acidified lake, including an increase in water transparency and resulting changes in the thermal characteristics of the lake.

Dinoflagellates do not always dominate in acidic lakes. In several Florida lakes with pHs between 4.5 and 5.5, greens represent about 60% of the total plankton, whereas blue-green populations are low.[27] Kwaiatkowski and Roff[22] observed that the relative abundance of greens and blue-greens in the Sudbury area may be closely related with pH. In circumneutral lakes, greens formed 40 to 50% and blue-greens 30% of the total algal flora, whereas in acidic lakes blue-greens formed 60% and greens only about 25% of the total algal assemblage. The dominance of blue-greens in acidic lakes was largely due to blooms of *Merismopedia* sp. Similarly, in Florence Lake (Sudbury) and in some Swedish lakes, blue-greens comprise more than half of the total phytoplanktonic community, despite the very acidic nature of these lakes (pH <4.5).[28,29] As in the previous examples, blooms of *Merismopedia* sp. were responsible for the blue-green dominance. The abundance of blue-greens in acidic lakes appears to contradict some earlier physiological studies that suggest blue-greens should be outcompeted in waters with pH <4.8.[30] A complex of factors undoubtedly determines the outcome of competition between algae.

The low abundance of chrysophytes in several of the well-studied acidic lake regions,[15,16,27,31] as well as some of the studies cited above, might suggest that chrysophytes are poor competitors at low pH. However, this again is not universally true. In fact, a collation of phytoplankton data from many acid-stressed regions indicates that chrysophytes may continue to be important primary producers, and discrepancies between lake regions may be due to differing nutrient levels,[32] or in some cases simply to differences in sampling procedures.[9] For example, in a survey of 56 northern Ontario lakes, chrysophytes dominated in every pH interval.[9] Similarly, chrysophytes dominated, both in terms of species number and biomass, in acidic high mountain lakes in Switzerland,[33] in two Adirondack Mountain lakes,[34] and in Wavy Lake, Sudbury (pH <5) where *Dinobryon* was the dominant genus.[28]

The presence of certain algal assemblages in acidic lakes indicates their physiological ability to tolerate low pH and/or variables closely associated with pH; however, relatively little is known about the mechanisms involved. Ion transport across the cell membrane may be influenced by low pH. During the study of H^+ ion tolerances of *Chlamydomonas acidophila* and *C. sphagnophila*, Cassin[18] found that H^+ ion does not appear to enter at low pH, suggesting that a cell boundary mechanism may be involved in their acid tolerance. Species of *Chlorella*, *Chroococcus*, and *Gymnodinium* secrete thick gelatinous sheaths, which may form partial barriers to the acidic environment, thus providing a competitive advantage.[26]

An important environmental consequence of acid rain is the increased input of soluble metals, both from the lake drainage and from direct atmospheric deposition. In clearwater acidic lakes, metal concentrations can be several orders of magnitude higher than in circumneutral and alkaline waters.[35,36] High metal concentrations can be toxic to algae, and may therefore be important factors influencing phytoplankton distribution. Moreover, certain taxa may develop mechanisms to sustain their populations in low pH, high trace metal concentration waters.[3,16,37]

Unfortunately, little comprehensive data are available on the effects of specific metals on algal taxa, especially since few field studies have separated the effects of metals from pH. As a general rule, blue-greens and diatoms are less tolerant to metals than greens.[38] Increases in hydrogen ion concentrations can affect algal cells directly by affecting uptake mechanisms and depressing photosynthesis,[39] or indirectly by regulating the speciation of the total metal pool.[38] At low pH, algal metal uptake may be lowered or may completely stop.[40] In a laboratory experiment, Peterson et al.[41] found that free metal ions became more toxic to *Scenedesmus quadricauda* when H^+ ion concentrations declined. Gensemer and Kilham[42] observed that growth rates of *Asterionella formosa*, *Stephanodiscus hantzschii*, *Scenedesmus* sp., and *Anabaena flos-aquae* greatly declined when the pH was lowered below 6.0. However, the growth was partially or completely restored when the trace metal concentrations were lowered in the growth media.

Elevated metal concentrations, within limits, may also enhance the growth of certain algae. For example, *Chrysochromulina breviturrita*, a prymnesiophycean alga that often forms unpredictable blooms in soft-water lakes,[24] requires selenium,[43] an element that is closely related with atmospheric sulfate deposition.[44] Clearly, metal/pH/algal relationships are complex, and more empirical data are needed.

B. Species Richness and Diversity

Phytoplankton species richness is generally lower in acidic lakes.[14-16,22] Yan[16] recorded only half of the phytoplankton genera in lakes impacted by acidic precipitation than in morphometrically similar unaffected lakes. Maximum species richness changes have been identified between pH 5.0 and 6.0.[14,16] In a study of 115 acid-sensitive lakes on the Swedish west coast, Almer et al.[11] found that lakes with lower pH (<5.0) contained a phytoplankton community consisting of about 10 species, and the very acidic lakes (pH <4) contained only 3 taxa. Meanwhile, in lakes with pH 6 to 8, 30 to 80 phytoplankton species were present. Similarly, in the metal-contaminated acidic lakes of Sudbury, a marked decline in species richness was observed between pH 7.15 and 4.05 in all major algal groups.[22] The number of green algal taxa declined from 26 to 5, chrysophytes from 22 to 5, and blue-greens from 22 to 10 species. In acidic Norwegian lakes (pH <5) similar declines in species diversity were observed,[45] even though the total number of species occurring over an entire season may actually be higher. This may be a possible explanation why in Lake 223 no significant changes were observed between year to year averages of phytoplankton species richness and diversity.[25]

Algal diversity, like other observations discussed in this paper, cannot simply be related to pH, and exceptions always occur. For example, in acidic brownwater lakes, algal diversity is often higher than in both clearwater acidic and nonacidic lakes.[45,46] This may be because in humic lakes seasonal succession of algal species occurs at a higher rate.[45]

C. Biomass and Production

Little quantitative data are yet available to explain the impact of acidification on algal biomass and production. Algal biomass and production will be influenced greatly by seasonal succession, the tolerance of species to changing water quality, nutrient availability, and light penetration.[47] These factors may all be influenced by changes in lakewater acidity.

In metal-contaminated Sudbury lakes, Kwiatkowski and Roff[22] reported increased biomass with decreases in pH, whereas in some Ontario lakes Hendrey et al.[3] found that phytoplankton biomass in acidic lakes was significantly less than in circumneutral lakes. In enclosure studies, Yan and Stokes[15] recorded an increase in biomass when pH was lowered to 5.1 in an acidic lake, whilst Marmorek[48] observed no change in algal biomass in his enclosure study.

In Lake 223 no significant changes were observed in phytoplankton biomass when the pH of the lake was lowered from 6.8 to 6.1.[25] Once the pH reached 5.6, phytoplankton biomass increased and remained high even when the lake pH dropped to 5.1. Since the biomass increase was most pronounced in the hypolimnion, increased clarity in the upper waters may have been an important factor.[49] Moreover, the replacement of many smaller species by larger taxa may have also played an important role.[26]

Studies have suggested that phytoplankton biomass is generally less than 1 mg/l in acid lakes.[14,16,22] Although over an entire year the phytoplankton production in Lake Gårdsjön remained less than 1 mg/l, seasonal variations (0.01 to 0.75 mg/l) were large.[50] Low biomass was observed between February to April and July to August. The highest biomass was observed during the spring bloom in late May and during a small bloom in early September. Yan[16] also cautioned that in acidic as well as nonacidic lakes algal biomass often varies greatly between sampling dates, further stressing the difficulties associated with small sample sizes.

In Sudbury lakes, although algal biomass remains relatively high in acidic sites (pH <5.0), primary production is significantly lower than in nonacidic lakes.[22] Similarly, in the Turkey Lakes watershed primary production was lower in the poorly buffered than in the well-buffered lakes.[51] The comprehensive review of the National Research Council of Canada[52] documented that primary production for the majority of acidic Canadian Shield lakes ($\sim$3 mg C m^{-3} h^{-1}) is within the range reported for nonacidic lakes (0.3 to 6.9 mg C m^{-3} h^{-1}). The primary production (0.3 to 2.9 mg C m^{-3} h^{-1}) in acidic Lake Gårdsjön[53] was also within this range.

The relationship between phytoplankton biomass and pH is still controversial, but there is growing evidence that the pH effect is not a direct one. Other factors, such as inorganic nutrients and water clarity, are likely to be more important. For example, studies indicate that algal biomass, primary production, and chlorophyll *a* concentrations can be better explained by lakewater phosphorus concentrations than by pH.[52] For lakes ranging in pH from acidic to circumneutral, Yan[16] found correlations between total phosphorus and biomass but no correlation with pH. In the Sudbury environmental study,[54] Middle Lake was artificially neutralized from a pH of 4.4 to 7.0, yet algal biomass did not increase until phosphorus concentrations were also raised. Similarly, the addition of phosphorus to enclosures of an acidic lake (pH 5.0) resulted in an immediate algal bloom.[55]

Phosphorus availability is not independent of pH and/or pH-related factors. Aluminum, whose solubility is closely related to pH, plays an important role in the complexation of phosphorus.[11] In clearwater acidic lakes aluminum concentrations can be as much as two orders of magnitude higher than in circumneutral lakes.[36] The affinity of Al for P results in the precipitation of aluminum phosphate. The removal of phosphorus by this mechanism is greatest between a pH of 5 and 6.[56] This may explain why Almer et al.[11] found lower biomass in lakes of pH between 5.1 and 5.6. From a laboratory study, Pillsbury and Kingston[57] indicated that the dominance of *Asterionella ralfsii* var. *americana*, *Arthrodesmus indentatus*, *A. octocornus*, *A. quiriferus*, *Staurastrum arachne* var. *curvatum*, *S. longipes* var. *contractum*, and *S. pentacerum* decline in acidic waters (pH <5.7) containing high aluminum concentrations (>50 μg/l). It is possible that these taxa may be indicators of acidic waters of low monomeric aluminum. As a compensatory mechanism for the low availability of phosphorus in some acidic lakes, certain phytoplankton species (e.g., *Chlamydomonas*, *Chromulina* spp.) may produce large quantities of phosphatase.[58] High aluminum concentrations may induce phosphatase activity.[59]

The availability of other nutrients will also be affected by pH and pH-related factors. For example, the amount of total inorganic carbon declines as pH decreases,[60] and in soft-water lakes dissolved inorganic carbon may become growth limiting. Turner et al.[61] indicated that net photosynthesis is strongly correlated with dissolved inorganic carbon (DIC), and because DIC is related to pH, inorganic carbon may limit the growth of blue-greens in acidic waters. Moreover, the form and amount of inorganic carbon available in acidic lakes may elicit a different response in filamentous blue-greens and greens. Although these carbon deficiencies may stop photosynthesis for a large percentage of a day,[62] overall production in acidic lakes is not usually carbon limited because sources such as atmospheric diffusion, bacterial activity, and benthic respiration replenish the CO_2 pool.[16,21,63] Experimental evidence indicates, however, that differences in DIC in acidified lakes may influence the composition of phytoplankton assemblages.[64]

D. Herbivore-Algal Interactions

Although studies dealing with herbivore-algal interactions in acidic lakes are rare, we believe that this topic deserves further emphasis. In acid-stressed lakes, zooplankton communities tend to be simple with low species diversity. *Bosmina longirostris* frequently dominates.[65,66] Because herbivorous zooplankton are often selective grazers, changes in herbivore pressures may exert important influences on the algal community. For example, dinoflagellates such as *Peridinium* spp., which may contribute 30 to 40% of the phytoplankton biomass,[16] are not readily ingested by herbivores.[67] Eriksson et al.[19] suggested that the dominance of dinoflagellates in acidic lakes may be closely related to zooplankton grazing, but other factors must again be considered. For example, Havens and DeCosta[68] experimentally showed that the dominance of *P. inconspicuum* in their study was primarily due to its tolerance for low pH. Singer et al.[23] concluded that zooplankton herbivory can influence phytoplankton assemblage and production, but not completely control them.

These studies indicate that, although algal biomass may remain high in acidic lakes, a sizable portion of the phytoplankton may be inedible.[26] This may have a considerable impact on energy transfer efficiencies.[66]

III. BENTHIC ALGAL COMMUNITIES OF ACIDIC LAKES

Benthic algae are found in close association with aquatic plants, animals, and other substrates such as sediments, rocks, and organic debris. In lakes, the occurrence of benthic algae is commonly restricted to shallow waters. However, with acidification, phytoplanktonic communities often diminish in size, thus increasing the photic zone.[22,27] Increased water clarity may also be due to (1) the precipitation of humic substances as a result of complexation with aluminum,[11,56] (2) a change in color of the dissolved material as pH decreases,[69] (3) a change in phytoplankton composition to fewer but larger species,[15] and (4) the dissolution of iron and manganese colloids.[70] Consequently, light penetrates deeper in acidified lakes, and the importance of benthic algae increases. Because the response time of benthic algal communities to changes in pH is less than 2 weeks,[71] and because these growths are often easily noticed, they may be useful predictors of lakewater acidity changes.

Experimental data confirm observations that benthic algae often thrive in acidified lakes. For example, in stream experiments the growth of periphytic algae was 40% higher in acidic (pH ~4) than in nonacidified waters.[3] Growth continued to increase as the pH was lowered. Although in this and other experimental studies[72] attempts were made to stimulate the natural environment, their applicability may still be in question because often the pH was dropped rapidly, the observations were made over short time periods, and the implications of other biotic and abiotic factors were not assessed.

Although benthic algal abundance often increases in acidic waters, species richness and diversity often decline significantly.[73,74] The most visual impact of acid precipitation on poorly buffered lakes is the increase in the growth of filamentous algae. The formation of an algal mat contributes to simplification of algal community composition (as many of these mats are unialgal), detrimentally alters habitat for other organisms,[25] and may inhibit microbial decomposition, which in turn contributes to the retention of nutrients in living and unmineralized plant matter.[75]

Because the development of filamentous algae usually accelerates at about pH 6, their abundance may be a very useful indicator of early acidification.[61] The algal mats can be completely eliminated from acidic lakes by raising the pH above 6.0.[76] The species richness tends to decline in algal mats as the waters acidify.[77]

Three major types of algal mats can usually be distinguished: (1) felt-like mats dominated by blue-greens, (2) mats composed of less cohesive green algae, especially members of the Zygnemataceae, and (3) loosely packed greens commonly attached to aquatic plants.[77]

In Lake Gårdsjön (Sweden) the algal mat was type 1, densely constructed of motile filaments of *Lyngbya burrellyana (Phormidium ambiguum)* with less abundant filaments of *Oscillatoria geminata.*[78] Other benthic algae (*Merismopedia* spp., *Aphanothece* spp., *Frustulia rhomboides,* and *Anomoneoeis serians*) were also abundant. Similarly, in Lake Colden (New York) the algal mat was primarily composed of the blue-green alga *Phormidium tenue,* with lesser amounts of *Mougeotia, Tabellaria fenestrata, Fragilaria virescens,* and *Diatoma.*[75]

As noted previously with the phytoplankton, the abundance of blue-greens in algal mats does not follow Brock's[30] observations that these algae are absent from waters of pH 5 and lower. Lazarek[78] suggested that, although blue-greens are sensitive to low pH, species belonging to the Oscillatoriaceae can withstand acidic waters by an increase in mucilage production. Because this substance is mainly composed of polysaccharides, it can chelate toxic metals. However, Turner et al.[61] observed that filamentous blue greens (i.e., *Lyngbya* and *Anabaena*) undergo metabolic stress in acidic waters as indicated by lower primary production.

In the experimentally acidified Lake 223 the algal mat was primarily of type 2. The abundance of blue-greens in this lake greatly declined with acidification, whereas greens

(*Mougeotia* spp.) became dominant as the pH was lowered below 6.0 (Turner, M., personal communication).[61] Although *Mougeotia* invaded as pH fell below 6.0,[61] lab experiments have suggested an optimum pH of 4.5 for this taxon.[76] Variations may exist between different strains.[79]

Factors responsible for the growth of algal mats are not well understood. The most common explanation is that in acidic lakes filamentous algae simply predominate because their physiological optima occur in low pH waters. The occurrence of algal mats is generally most luxuriant in the upper littoral zone, thereby reflecting the need for high light for optimum growth. Moreover, water movements in this region may benefit the greens because the thickness of the cell membrane boundary layer decreases, thus increasing the potential for nutrient uptake.[61] This would provide an added advantage to filamentous algae, which have added adaptations for sequestering nutrients in low light environments.[80]

Desmids are an important group of nonmotile green algae, whose habitat characterization still represents a problem. These algae are occasionally reported in the plankton, although many are benthic or tychoplanktonic, and they are therefore discussed here. In general, brownwater acidic lakes support diverse desmid population, but no single species predominates. Desmids are characteristic of relatively unpolluted waters of low mineral contents with a slightly acidic pH.[7] Taxa such as *Cylindrocystis brebissonii, Bambusina brebissonii, Closterium striolatum, C. directum,* and *Staurastrum jaculiferum* may be very common in strongly acidified waters.[47] Their abundance in acidic waters, however, is not universally high. In a long-term study of the algae of some Dutch moorland pools, desmid abundance dropped off strikingly with acidification.[81] In one of the studied regions, desmid species richness declined from a high of 195 in 1925, to 123 in 1955, and then to only 68 in 1975.

IV. DIATOMS AND LAKEWATER pH

A. Diatoms and Lakewater pH

Diatoms are often important components of algal assemblages in lakes, yet an examination of our discussion thus far might suggest that they provide relatively little information on acidification trends, except that planktonic forms often decline,[14] whereas in benthic habitats diatoms appear insignificant among the more visible blue-green and green algal mats. Nonetheless, diatoms provide important information on acidification trends — not from cursory examinations of net tows or water samples, but from critical, high resolution microscopic observations of acid-cleaned cells.[82]

An important characteristic of diatoms is that these single-celled algae are encased in a siliceous cell wall or frustule, which is made of two interlocking halves or valves. The size, shape, and sculpturing of the frustule is species specific. In fact, the entire systematics of diatoms is based on frustule morphology. Moreover, because they are siliceous, diatom valves are usually well preserved in lake sediments. The resulting stratigraphic studies of sediment cores, showing diatom changes at the species and subspecies level with presumed acidification, have brought diatom taxonomy and ecology to the forefront of many lake acidification studies.[83-86]

Diatoms thrive in most aquatic environments, including the open plankton, and often diverse assemblages attached to substrates, such as higher plants (epiphytic), mud (epipelic), rocks (epilithic), and sand (epipsammic). One of the most efficient methods of studying diatom distributions with respect to lakewater conditions is to identify and enumerate the frustules preserved in the surface sediment (e.g., top centimeter) of study lakes. It is assumed that the top centimeter contains the last few years of sediment accumulation. These surface sediments integrate diatom populations both in space, from the many habitats within the lake, and in time over several years. From these data, autecological and synecological information can be gleaned.

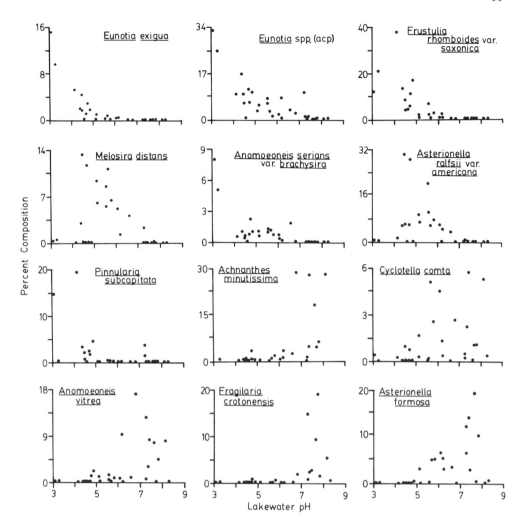

FIGURE 1. The percent composition of selected diatom taxa, identified from the surface sediments of 30 Sudbury lakes, plotted relative to the lakewater pH.

Surface sediment studies (so-called ''calibration sets'') have repeatedly shown that over a diverse range of lakewater characteristics, the distribution of most diatoms is significantly correlated with pH and/or pH-related variables.[87-90] Dramatic floristic changes occur as lakewater acidity increases (Figure 1). In fact the correlation with pH is usually higher than for any other variable.

The historical development of diatom-based pH reconstructions has been reviewed by Battarbee et al.[91] The recognition that pH had an important influence on diatom distribution came early in diatom work, when Hustedt[92] delineated five pH categories, namely:

1. Acidobiontic: diatoms occurring at pH of about 5.5 or less
2. Acidophilous: occurring at pH about 7, with the widest distribution at pH less than 7
3. Indifferent: equal occurrence on both sides of pH 7
4. Alkaliphilous: occurring at pH value about 7, with widest distribution at pH above 7
5. Alkalibiontic: occurring at pH above 7

In surface sediment studies, Hustedt's pH categories are often still used today, with minor terminological changes.[93-95] Because Hustedt assigned the indifferent category to diatoms that are equally common on both sides of pH 7, the term indifferent has been substituted by circumneutral in some studies and the indifferent category is assigned to diatoms that have no specific pH preference. The grouping of diatom species to any pH category is usually based on published ecological information and on their known distribution in the study area.

Surface sediment studies have shown that certain diatoms are common to specific pH regimes. For example, common acidobiontic diatoms include *Actinella punctata*, *Anomoeoneis serians*, *Eunotia exigua*, *Navicula subtilissima*, *Semiorbis hemicyclus*, *Tabellaria binalis*, and *T. quadriseptata*.[87,89,90,93,96-99] All these acidobiontic taxa are benthic forms. Although a majority of these acidobionts may be present in many acid lake regions, the dominants usually vary between regions. For example, in metal-contaminated Sudbury lakes *E. exiqua* and *T. quadriseptata* dominated,[89] whereas in Algoma lakes dominant forms were *A. punctata*, *A. serians*, *S. hemicyclus*, and *T. binalis*.[98] Battarbee[83] pointed out that the restriction of acidobionts in waters with pH less than 5.5 suggests that they are most competetive in acidic environments where bicarbonate alkalinity is not present. In very acidic lakes, diatoms may be very abundant, but species diversity may be low.[93,100] Although acidobionts are generally restricted to very acidic waters, a major portion of the diatom community of these lakes is often comprised of acidophilous taxa that have wider pH tolerances. For example, in some extremely acidic (pH <3.5) Sudbury lakes, acidophilous taxa *Frustulia rhomboides* var. *saxonica*, *Anomoeoneis serians* var. *brachysira*, *Pinnularia subcapitata*, and many acidophilous *Eunotia* species made up to 60% of the total diatom assemblage.[101] Similar observations have been reported by Patrick,[102] Bennett,[103] and Brugam and Lusk.[104] In very acidic waters, alkaline diatom taxa are generally absent or present in only trace amounts.

In comparison to acidobiontic taxa, a much larger number of diatom species are categorized as acidophilous. Although Hustedt delineated the acidophilous category as those diatoms that are most common in waters of pH less than 7, surface sediment studies have indicated that these taxa are even more restrictive in their distribution, and often predominate at pH 6 and lower.[96-100] To some extent this depends on the distribution of lake pH in a regional set. Commonly reported acidophilous diatom taxa are *A. serians* var. *brachysira*, *Asterionella ralfsii* var. *americana*, *Achnanthes marginulata*, many *Eunotia* species, *Frustulia rhomboides* var. *saxonica*, *Melosira distans*, *M. lirata*, *Navicula bremensis*, *N. heimansii* (now called *N. leptostriata*), *N. mediocris*, *Neidium affine*, *P. biceps*, *P. hilseana*, *P. subcapitata*, *Surirella delicatissima*, and *Tabellaria flocculosa*.[87,89,90,96-100] In addition to pH, some of these taxa (e.g., *A. serians* var. *brachysira* and *F. rhomboides*) are sensitive to total organic matter concentrations in acidic lakes.[105]

In diatom studies, the indifferent and circumneutral categories have often been used interchangeably. Although regional differences occur, some of the frequently reported circumneutral taxa are *Achnanthes minutissima*, *Cyclotella comta*, *C. ocellata*, *C. stelligera*, *Navicula radiosa*, *N. pupula*, *Stauroneis phoenicenteron*, and *Synedra rumpens*.[87,89,90,96-100] The highest diatom diversity is often reported from circumneutral waters.[106] However, in sedimentary diatom studies the use of species richness and diversity should be avoided because differing sedimentation rates will introduce artifacts into the calculations.[107]

In alkaline lakes (pH >7), alkaliphilous diatoms predominate, with lesser numbers of circumneutral taxa. Common alkaliphilous diatoms are *Achnanthes linearis*, *Anomoeoneis vitrea*, *Asterionella formosa*, *Cyclotella michiganiana*, *Fragilaria brevistriata*, *F. crotonensis*, *F. construens* var. *venter*, *F. pinnata*, *Melosira ambiqua*, and *Nitzschia gracilis*.[87,89,90,96-100] Alkalibiontic diatoms are not found in soft-water lakes.

Several workers have observed that planktonic diatoms disappear when pH declines below 5.6,[14,87] and the loss of planktonic diatoms is the first major sign of lake acidification.[83] However, this hypothesis requires a careful evaluation for individual study regions because,

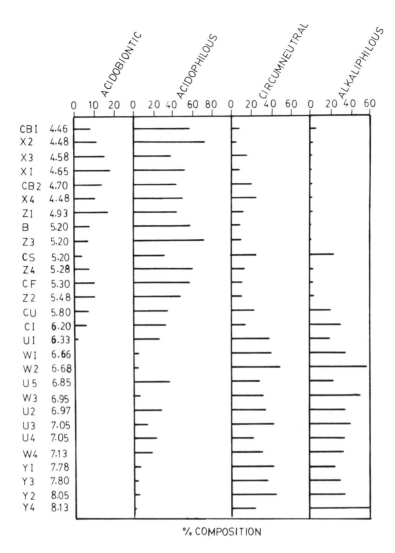

FIGURE 2.   The percent composition of pH indicator diatom assemblages from the surface sediments of 28 Algoma, Ontario lakes. (From Dixit, S. S., *Can. J. Bot.*, 64, 1129, 1986.)

for example, in acidic (pH <5.6) Ontario and Quebec lakes, planktonic diatoms were present in high abundance.[89,108,109] Water depth has been positively correlated with abundance of planktonic diatom species in these lakes.[108,110]

With the exception of extremely acidic waters (pH <4.0), common taxa generally contribute <20% of the total diatoms identified in surface sediment, thus emphasizing the importance of rare taxa. The assignment of these many taxa (perhaps over 100 species in a single diatom count of 500 valves) to Hustedt's pH categories is therefore a practical and efficient way of including all the diatom information in pH interpretations.[87,101,111] Significant relationships (Figure 2) can be shown between lakewater pH and the acidobiontic, acidophilous, circumneutral, and alkaliphilous assemblages.

B. Diatom-Inferred pH Calibration

The relationship between the diatom assemblage in the surface sediment of a lake and lakewater pH can be quantified or calibrated to a surprising degree of accuracy. Partly due to regional differences, these calibration relationships are not readily transferable from one area to another,[111,112] thereby requiring a separate calibration for each lake ragion. Although many factors may be responsible for this, the lack of comparable data between studies requires special attention.

Using Hustedt's pH categories, Nygaard[113] pioneered the quantification of diatom-inferred lakewater pH during his classic study of Lake Gribsø. He recognized that acidobiontic and alkalibiontic species are better ecological indicators (i.e., their pH preference is narrower) than acidophilous and alkaliphilous diatoms. Consequently, he determined that the abundance of acidobiontic and alkalibiontic diatoms should be more heavily weighted in predictive functions. Three indices were developed:

1.   Index α = (acid units)/(alkaline units)
2.   Index ω = (acid units)/(number of acidic taxa)
3.   Index ϵ = (alkaline units)/(number of alkaline taxa)

The acid and alkaline units are computed by multiplying the percent acidobiontic and percent alkalibiontic species by ''5'' and then adding the product to the percent acidophilous and percent alkaliphilous diatom assemblages, respectively.

The science of quantifying lakewater pH to diatom assemblages took another leap forward when Meriläinen,[96] working on the surface diatom assemblages of 14 Finnish lakes, recognized that a significant, linear relationship can be shown between log index α and measured pH. Although Meriläinen[96] only claimed an accuracy of about one pH unit, recent studies have produced much better predictions.[87,89,95]

Among Nygaard's three indices, index α is repeatedly shown to be the best predictor of pH.[87,111,114] Nonetheless, the use of index α in acid waters may be problematic because of the poor representation of alkaline diatoms. A small shift in alkaline units resulting from the rare, chance occurrence of an alkaline taxon, a misidentification, or a wrong category assignment can seriously influence predicted pH.[87] To overcome this deficiency, Renberg and Hellberg[94] proposed a new index based on all of Hustedt's pH indicator assemblages.

$$\text{Index B} = \frac{\%\text{IND} + (5 \times \%\text{ACP}) + (40 \times \%\text{ACB})}{\%\text{IND} + (3.5 \times \%\text{ALP}) + (108 \times \%\text{ALB})}$$

Since its introduction, index B calibration has been used widely both in North America[87,89,95,105] and Europe.[94,115,116] However, similar to index α, one of the problems with index B is the weighting of various pH indicator assemblages. Although the various coefficients used in index B stabilize the index and reduce the variability in a data set, their ecological significance must be defined clearly for each study region. More recently alternative methods have been developed to predict lakewater pH. These include the use of multiple regression analysis of various pH indicator diatom assemblages, selected pH indicator diatom taxa, principal components of diatom taxa, and taxa clusters with measured lakewater pH.[87,89,95,111] The principle advantage in using multiple regression methods over indices is that they do not incorporate arbitrary coefficients, and, not surprisingly, multiple regressions of pH indicator assemblages are often the best predictors of lakewater pH (Figure 3), with an accuracy of less than 0.4 of a pH unit.[87,95,111] This method is also useful in predicting lakewater pH in metal-contaminated lakes.[89] However, recently objections have been raised to the linear assumptions of regression, and new techniques are being developed based on Gaussian distributions of species abundance.[117,118]

The surface sediment diatom-pH quantification for soft-water lakes has been established for various lake regions in North America and Europe, including the Adirondack Mountains, NY,[87] New England,[111] northern Great Lakes Region,[119] Ontario,[89,95,108] Quebec,[109,120] Atlantic Canada,[121] Scotland,[88] Finland,[96,122] Norway,[111] Sweden,[94] and West Germany.[123]

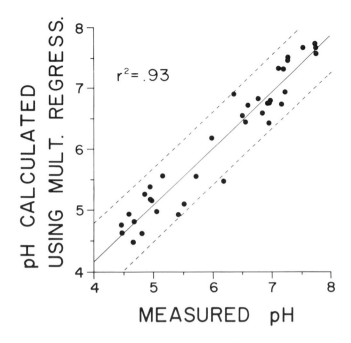

FIGURE 3.   The relationship between diatom-inferred pH using multiple linear regression of pH indicator diatom assemblages and measured lakewater pH of 37 Adirondack lakes. (From Charles, D. F., *Ecology*, 66, 994, 1985. With permission.)

C. Utility of Diatoms to Predict Recent Lakewater pH Changes

Many chemical, physical, and biological changes occur in lakes as a result of anthropogenic activity. Unfortunately, historical limnological data are often lacking and therefore paleo-limnological techniques are being used increasingly as the best available method to reconstruct past lake conditions. Contained in the sedimentary profile of a lake are a surprisingly large number of indicators, including geochemical markers, pigment degradation products, and the morphological remains of aquatic organisms. Diatoms, because of their well-preserved siliceous frustules and high ecological diversity, often form the mainstay of these studies.

Traditionally, fossil diatoms were used to interpret changes in lake trophic status resulting from human activity or from natural processes.[124-127] Not surprisingly, most recent work has centered on inferring pH changes in lakes that are suspected of having acidified recently.[85] The standard approach is to remove a short (approximately 50 cm long) sediment core from near the deepest part of the study lake, section the sediment at close intervals (e.g., 1-cm), remove the organic matrix using concentrated acids, and identify and enumerate the diatoms from each stratigraphic level. The pH inference transfer functions, constructed from a suitable calibration set (see above), can then be applied to the core data and a diatom-inferred pH profile can be calculated.[82] Because recent sediments can be dated using ^{210}Pb chronology,[128,129] the timing and rates of acidification can also be estimated.

Diatom studies have documented the rates and magnitude of acidification for selected lakes in the Adirondack Mountains,[130,131] New England,[132] the Midwest and Florida,[133] Ontario,[112,114,133] Sweden,[94,135] Finland,[115] Norway,[97,111] Scotland,[88,116] West Germany,[136] and The Netherlands.[137]

D. Recent Acidification in Clearwater Lake: A Case Study

Clearwater Lake (46° 22′N 81° 03′W) is located 12 km southwest of Sudbury, an area where metal smelting has been underway since the latter part of the 18th century.[138,139] As early as the 1950s, Sudbury area lakes were recognized to be acid stressed[140] — the first in Canada. However, the timing, extent, and rate of acidification could not be addressed because long-term pH records were not available. A paleolimnological study was therefore initiated.[101]

Sedimentary diatoms were identified and enumerated from an 18-cm long sediment core from Clearwater Lake.[134] The core was dated using ^{210}Pb chronology. The downcore distribution of common diatoms indicated that prior to 1930 the diatom population remained unchanged; however, since that time major shifts occurred (Figure 4), with the relative abundance of acidobiontic taxa (*Tabellaria quadriseptata* and *Eunotia exiqua*) increasing in the more recent sediments. Dramatic increases were also recorded in many acidophilous taxa (*Frustulia rhomboides* var. *saxonica*, *Pinnularia subcapitata*, *P. sudetica*, *Eunotia* sp., *Surirella delicatissima*, *Neidium iridis* var. *amphigomphus*, and *Navicula mediocris*).

The multiple regression equation developed for a diatom calibration set of 30 Sudbury lakes[101] was then applied to the Clearwater core.[134] The resulting inferred pH profile indicates that prior to 1930, the pH of the lake ranged between 6.0 and 6.5, but since then there was a very rapid decline in lakewater pH from a high of about 6.0 to a low of 4.2 (Figure 5). The drop in lakewater pH lasted about 40 years. However, since about 1970, the pH decline had ceased in Clearwater Lake. This stratigraphic sequence corresponds closely to the relatively unchanged pH values that are available for this site between 1973 and 1984,[134] and is a reflection of the recent decline of the smelting industry in Sudbury.

The above pH reconstruction provides strong evidence that Clearwater Lake had acidified rapidly as a result of increased SO_2 emissions. For lakes located close to the smelters, the pH decline may have commenced soon after the open pit roasting of ore started;[101] however, in lakes more distant from the point sources (e.g., Clearwater Lake), the acidification only began after the installation of tall stacks in the late 1920s.

V. CHRYSOPHYTES AND LAKEWATER pH

Chrysophytes in the family Mallomonadaceae, which include important genera such as *Mallomonas* and *Synura*, are flagellated euplanktors that are sensitive indicators of pH. Because these algae are covered by a siliceous armour of species-specific scales, they can be studied in sedimentary profiles using the same techniques developed for fossil diatoms.[141] Paleolimnological investigations using scaled chrysophytes lag behind that of diatoms, with research on the former beginning only in this decade.[142-144] Despite this relatively short time period, research has accelerated greatly.

Calibration sets for chrysophytes are now available for lakes in the Adirondacks,[145] the Sudbury region,[101] Quebec,[146] and Finland,[147] as well as ongoing studies in our lab for lakes in New England, southern Ontario, and Norway. Transfer functions relating chrysophyte distributions to lakewater pH are only now being developed; nevertheless striking relationships can be shown between species abundance and pH. For example, data from a 38-lake calibration in the Adirondacks[141,145] clearly showed that taxa such as *Synura echinulata*, *Mallomonas hamata*, and *M. hindonii* are more common in acidic waters, while *S. spinosa*, *M. caudata*, and *M. punctifera* are more competitive in alkaline waters (Figure 6). Cluster analysis and reciprocal averaging of the data also indicated that pH and pH-related factors had an overriding influence on the composition of assemblages.[145]

It was not long before chrysophyte scales were included in paleolimnological studies of lake acidification. Smol et al.[148] showed that acidobiontic species, such as *M. hindonii* and *M. hamata*, increased strikingly in the recent sediments of presently acid (pH = 4.8) Deep Lake (Adirondacks). This shift coincided with the demise of a trout fishery, presumably due to acidification. Several other studies soon followed.[101,149-152]

As an example, we present the recent chrysophyte history of Big Moose Lake (43° 49′ 02″ N, 74° 51′ 23″ W), a presently acidic (pH = 4.6 to 4.9) Adirondack lake (Figure 7).

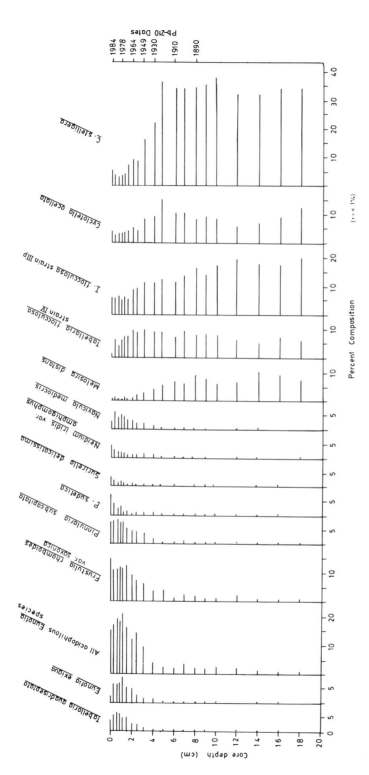

FIGURE 4. Relative abundance of common diatom taxa in a sediment core from Clearwater Lake. (From Dixit, S. S., Dixit, A. S., and Evans, R. D., *Sci. Total Environ.*, 67, 53, 1988. With permission.)

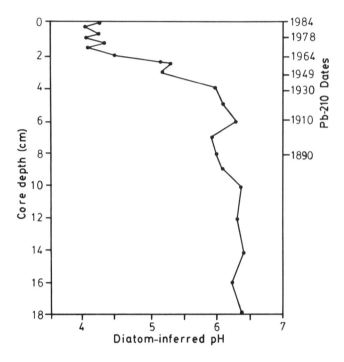

FIGURE 5.   Diatom-inferred pH profile from Clearwater Lake. (From Dixit, S. S., Dixit, A. S., and Evans, R. D., *Sci. Total Environ.*, 67, 53, 1988. With permission.)

We chose this site because it contrasts with the Clearwater Lake example (Figure 4) in that the Adirondacks are distant from point sources of sulfuric acid emissions, yet many of the lakes are known to be acidifying.[153] In addition, a wealth of ancillary paleolimnological data are available for Big Moose Lake.[154]

Prior to the early 1900s, the chrysophyte flora was dominated by *M. crassisquama* — a taxon that has a wide distribution in nonacidic waters. Near the turn of the century more acidophilous taxa, such as *M. acaroides, S. echinulata,* and *S. sphagnicola* increased in relative abundance. These species shifts indicate a modest acidification, most likely reflecting the early stages of atmospheric deposition of inorganic acids. Interestingly, at this site, as well as several others, this early change in the chrysophytes is not recorded in the diatom profiles for the same cores,[133,149,154] indicating that chrysophytes may be more sensitive markers of early acidification.

The increase in taxa that are abundant in only very acid waters, such as *M. hamata* and *M. hindonii,* clearly delineates the period of marked acidification in Big Moose Lake, a trend that is shown using many other paleoecological indicators.[154] For example, Charles[131] estimated that the diatom-inferred pH dropped about one pH unit from about 5.6 at the 7-cm level to 4.6 at the surface. This inferred pH shift coincides with historical records of fish declines, with acid-sensitive species being the most affected.

The above data show that chrysophytes are useful adjuncts to reconstructions of lakewater pH based on diatoms and other indicators, as they provide independent confirmatory evidence of overall trends. Equally, the data show that additional information, such as the timing of the early acidification phase, can be gleaned from a careful analysis of chrysophyte microfossils. The observation that spring snowmelt may be an especially stressful time for acid-sensitive organisms may be one reason why vernal blooming chrysophytes appear to be affected before diatom populations in some lakes. Chrysophytes may also be more sensitive to certain metals,[101,155,156] although these observations require further study (research in progress).

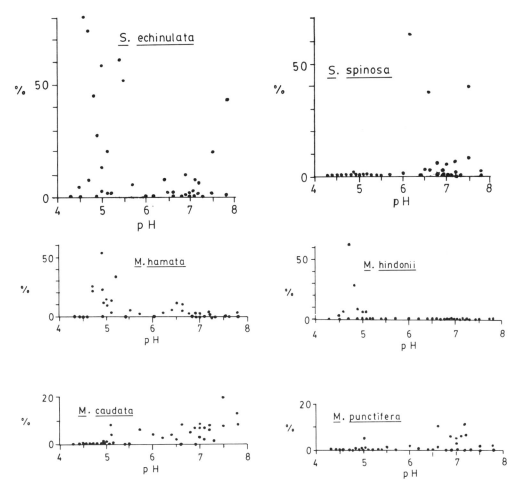

FIGURE 6. The percent composition of selected chrysophyte scales plotted relative to the measured lakewater pH of 38 Adirondack lakes. (From Smol, J. P., *Diatoms and Lake Acidity*, Smol, J. P., Battarbee, R. W., Davis, R. B., and Meriläinen, J., Eds., Dr. W. Junk, Dordrecht, Netherlands, 1986, 275. With permission.)

VI. SUMMARY AND CONCLUSIONS

We began this review noting that much descriptive data are available for algal populations in acid-stressed environments. Yet, we often conclude that more research will be required before any substantiated conclusions can be made. These seemingly paradoxical statements reflect, to some degree, methodological problems and shortcomings, the paucity of appropriate laboratory and lake manipulation studies, and, to a large degree, taxonomic imprecisions. The lack of long-term historical data further deters from any comprehensive overview, although the rapidly developing field of paleolimnology will undoubtedly continue to add important information. Nevertheless, the real dilemma rests with the very complex nature of algal assemblages — not only because so many taxa and life strategies are involved, but also by the many interrelated physical, chemical, and biological variables that influence the composition of communities. Synergistic and antagonistic reactions must also be considered. These factors are all related to a myriad of limnological changes that occur when lakewater pH is altered.

Some generalizations can be made about the distribution of algal assemblages in acid-stressed lakes. Among planktonic taxa *Merismopedia*, *Peridinum*, and *Gymnodinium* are common, whereas *Mougeotia*, *Zygnema*, *Phormidium*, and *Spirogyra* often dominate the

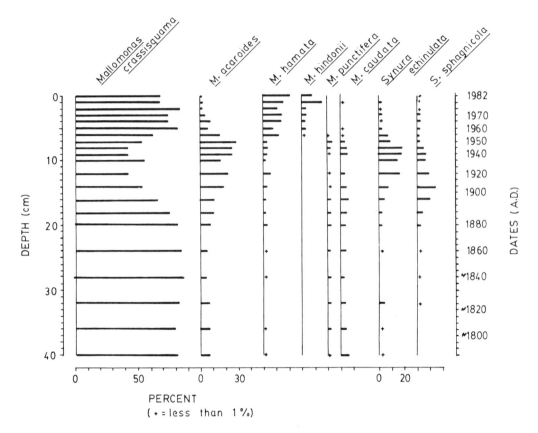

FIGURE 7. The relative abundance of chrysophyte scales in the recent sediments of Big Moose Lake. (From Smol, J. P., *Diatoms and Lake Acidity,* Smol, J. P., Battarbee, R. W., Davis, R. B., and Meriläinen, J., Eds., Dr. W. Junk, Dordrecht, Netherlands, 1986, 275. With permission.)

benthic community. As lakes acidify, planktonic algal production may decline, while benthic taxa, often forming thick algal mats, become more important primary producers. Changes in species richness and community composition appear to be more reliable markers of acidification than biomass.

The presence and abundance of many diatom and chrysophyte taxa is closely associated with lakewater pH. Acidobiontic and acidophilous taxa (*Tabellaria quadriseptata, Asterionella ralfsii* var. *americana, Frustulia rhomboides,* various *Eunotia* species, *Mallomonas hamata, M. hindonii, Synura echinulata,* and others) occur commonly in acidic lakes, whereas circumneutral and alkaline taxa (*A. formosa, Cyclotella ocellata, C. meneghiniana, M. caudata, M. pseudocoronata, S. curtispina,* and others) dominate in circumneutral and alkaline waters, respectively. Because of the specific pH indicator value of these siliceous algae, various quantitative models have been developed that calibrate assemblages to lakewater pH. These models are widely used to interpret the history of lake acidification in many regions of North America and Europe.

The one unequivocal point that remains is that algae are of the utmost importance in any lake system, and therefore continued work is justified.

ACKNOWLEDGMENTS

Research in our lab on acid-stressed environments is supported by grants from the Natural Sciences and Engineering Research Council of Canada, Ontario Ministry of the Environment, and the U.S. Environmental Protection Agency. We thank Dr. John C. Kingston for his critical reading of this review.

REFERENCES

1. **Wright, R. F. and Gjessing, E. T.,** Acid precipitation: changes in the chemical composition of lakes, *Ambio,* 5, 219, 1976.
2. **Likens, G. E., Wright, R. E., Galloway, J. N., and Butler, T. J.,** Acid rain, *Sci. Am.,* 241, 43, 1979.
3. **Hendrey, G. R., Baalsrud, K., Traaen, T. S., Laake, M., and Raddum, G.,** Acid precipitation: some hydrological changes, *Ambio,* 5, 224, 1976.
4. **McFee, W. W.,** Sensitivity ratings of soils to acid deposition: a review, *Environ. Exp. Bot.,* 23, 203, 1983.
5. **Dillon, P. J., Jefferies, D. S., Snyder, W., Reid, R., Yan, N. D., Evans, R. D., and Scheider, W. A.,** Acidic precipitation in south-central Ontario: recent observations, *J. Fish. Res. Board Can.,* 35, 809, 1978.
6. **Gorham, E. and McFee, W. W.,** Effects of acid deposition upon outputs from terristrial to aquatic ecosystems, in *Effects of Acid Precipitation on Terristrial Ecosystems,* Hutchinson, T. C. and Havas, M., Eds., Plenum Press, New York, 1980, 465.
7. **Lee, R. E.,** *Phycology,* Cambridge University Press, Cambridge, 1980.
8. **Sze, P.,** *A Biology of the Algae,* William C. Brown, Dubuque, IA, 1986.
9. **Kelso, J. R. M., Love, R. J., and Lipsit, J. H.,** Chemical and biological status of headwater lakes in the Sault Ste. Marie District, Ontario, in *Acid Precipitation — Effects on Ecological Systems,* D'Itri, F. M., Ed., Ann Arbor Science, Ann Arbor, MI, 1982, 165.
10. **Kalff, J. and Knoechel, R.,** Phytoplankton and their dynamics in oligotrophic and eutrophic lakes, *Annu. Rev. Ecol. Syst.,* 9, 475, 1978.
11. **Almer, B. W., Dickson, W., Ekström, C., and Hörnström, E.,** Sulfur pollution and the aquatic ecosystem, in *Sulfur in the Environment, Part II,* Nriagu, J. O., Ed., John Wiley & Sons, New York, 1978, 271.
12. **Schindler, D. W. and Holmgren, S. K.,** Primary production and phytoplankton in the Experimental Lakes Area, northwestern Ontario, and other low carbonate waters, and a liquid scintillation method for determining C^{14} activity in photosynthesis, *J. Fish. Res. Board Can.,* 28, 189, 1971.
13. **Duthie, H. C. and Ostrofsky, M. L.,** Plankton, chemistry, and physics of lakes in the Churchill Falls region of Labrador, *J. Fish. Res. Board Can.,* 31, 1105, 1974.
14. **Almer, B. W., Dickson, W., Ekström, C., Hörnström, E., and Miller, V.,** Effects of acidification on Swedish lakes, *Ambio,* 3, 30, 1974.
15. **Yan, N. D. and Stokes, P. M.,** Phytoplankton of an acidic lake, and its responses to experimental alterations of pH, *Environ. Conser.,* 5, 93, 1978.
16. **Yan, N. D.,** Phytoplankton communities of an acidified heavy metal contaminated lake near Sudbury, Ontario: 1973-1977, *Water Air Soil Pollut.,* 11, 43, 1979.
17. **Moss, B.,** The influence of environmental factors on the distribution of freshwater algae: an experimental study. II. The role of pH and carbon dioxide-bicarbonate system, *J. Ecol.,* 61, 157, 1973.
18. **Cassin, P. E.,** Isolation growth and physiology of acidophilic *Chlamydomonas, J. Phycol.,* 10, 439, 1974.
19. **Eriksson, M. O. G., Henrikson, L., Nilsson, B. I., Nyman, G., Oscarson, H. G., Stenson, A. E., and Larsson, K.,** Predator-prey relations, important for the biotic changes in the acidified lakes, *Ambio,* 9, 248, 1980.
20. **Stenson, J. A. C. and Oscarson, H. G.,** Crustacean zooplankton in the acidified Lake Gårdsjön system, *Ecol. Bull.,* 37, 224, 1985.
21. **Conway, H. L. and Hendrey, G. R.,** Ecological effects of acid precipitation on primary producers, in *Acid Precipitation — Effects on Ecological Systems,* D'Itri, F. M., Ed., Ann Arbor Science, Ann Arbor, MI, 1982, 277.
22. **Kwiatkowski, R. E. and Roff, J. C.,** Effects of acidity on the phytoplankton and primary productivity of selected northern Ontario lakes, *Can. J. Bot.,* 54, 2546, 1976.
23. **Singer, R., Evans, G. L., and Pratt, N. C.,** Phytoplankton limitation by phosphorus and zooplankton grazing in an acidic Adirondack Lake, *J. Freshwater Ecol.,* 2, 423, 1984.
24. **Nicholls, K. H., Beaver, J. L., and Estabrook, R. H.,** Lakewide odours in Ontario and New Hampshire caused by *Chrysochromulina breviturrita* Nich. (Prymnesiophyceae), *Hydrobiologia,* 96, 91, 1982.
25. **Schindler, D. W., Mills, K. H., Malley, D. F., Findlay, D. L., Shearer, J. A., Davies, I. J., Turner, M. A., Linsey, G. A., and Cruikshank, D. R.,** Long-term ecosystem stress: the effects of years of experimental acidification on a small lake, *Science,* 228, 1395, 1985.
26. **Findlay, D. L. and Kasian, S. E. M.,** Phytoplankton community responses to acidification of Lake 223, Experimental Lake Area, northeastern Ontario, *Water Air Soil Pollut.,* 30, 719, 1986.
27. **Chrisman, T. L., Schulze, R. L., Brezonik, P. L., and Bloom, S. A.,** Acid precipitation: the biotic response in Florida lakes, in *Ecological Impact of Acidification,* Drabløs, D. and Tollan, A., Eds., SNSF Project, Oslo, 1980, 296.

28. **Conroy, N., Hawley, K., Keller, W., and Lafrance, C.,** Influence of atmosphere on lakes in the Sudbury area, Proc. 1st Spec. Symp. Atmospheric Contribution to the Chemistry of Lake Waters, International Association for Great Lakes Research, 146, 1975.

29. **Hultberg, H. and Anderson, I.,** Liming of acidified lakes: induced long-term changes, *Water Air Soil Pollut.,* 18, 311, 1982.

30. **Brock, T.,** Lower pH limit for the existence of bluegreen algae: evolutionary and ecological implications, *Science,* 179, 480, 1973.

31. **Bleiwas, A. S. H., Stokes, P. M., and Olaveson, M. M.,** Six years of plankton studies in the La Cloche region of Ontario, with special reference to acidification, *Verh. Int. Ver. Limnol.,* 22, 332, 1984.

32. **Dillon, P. J., Yan, N. D., and Harvey, H. H.,** Acid deposition: effects on aquatic ecosystems, *Crit. Rev. Environ. Control,* 13, 167, 1984.

33. **Schanz, F.,** Chemical and ecological characteristics of five high mountain lakes near the Swiss National Park, *Verh. Int. Ver. Limnol.,* 22, 1066, 1984.

34. **Hendrey, G. R.,** Effects of acidification on aquatic primary producers and decomposers, in *Acid Rain/ Fisheries,* Johnson, R. E., Ed., American Fisheries Society, Bethesda, MD, 1982, 125.

35. **Nortan, S. A., Davis, R. B., and Brakke, D. F.,** Responses of the northern New England lakes to atmospheric inputs of acids and heavy metals, Land Water Res. Cent. Proj. Rep. A-048-ME, University of Maine, Orono, 1981.

36. **Cronan, C. S. and Schofield, C. L.,** Aluminum leaching response to acid precipitation: effects of high-elevation watersheds in the northeast, *Science,* 204, 304, 1979.

37. **Wood, J. M. and Wang, H. K.,** Strategies for microbial resistance to heavy metals, in *Chemical Processes in Lakes,* Stumm, W., Ed., John Wiley & Sons, New York, 1985, 81.

38. **Stokes, P. M.,** Responses of freshwater algae to metals, in *Progress in Phycological Research, Vol. 2,* Round, F. E. and Chapman, D. J., Eds., Elsevier, Amsterdam, 1983, 87.

39. **Nalewajko, C. and Paul, B.,** Effects of manipulations of aluminum concentrations and pH on phosphate uptake and photosynthesis of planktonic communities in two Precambrian Shield lakes, *Can. J. Fish. Aquat. Sci.,* 42, 1946, 1985.

40. **Campbell, P. G. C. and Stokes, P. M.,** Acidification and toxicity of metals to aquatic biota, *Can. J. Fish. Aquat. Sci.,* 42, 2034, 1985.

41. **Peterson, H. G., Healy, F. P., and Wagemann, R.,** Metal toxicity to algae: a highly pH dependent phenomenon. *Can. J. Fish. Aquat. Sci.,* 41, 974, 1984.

42. **Gensemer, R. W. and Kilham, S. S.,** Growth rates of five freshwater algae in well-buffered acidic media, *Can. J. Fish. Aquat. Sci.,* 41, 1240, 1984.

43. **Wehr, J. D. and Brown, L. M.,** Selenium requirement of bloom-forming planktonic alga from softwater and acidified lakes, *Can. J. Fish. Aquat. Sci.,* 42, 1783, 1985.

44. **Crocket, J. H. and Kabir, A.,** Geochemical pathway studies of heavy metals in lake sediments from the Sudbury-Temagami area, Ontario, *J. Great Lakes Res.,* 7, 455, 1981.

45. **Raddum, G. G., Hobaek, A., Lømsland, E. R., and Johnsen, T.,** Phytoplankton and zooplankton in acidified lakes in south Norway, in *Ecological Impact of Acidification,* Drabløs, D. and Tollan, A., Eds., SNSF Project, Oslo, 1980, 332.

46. **Blouin, A. C., Lane, P. A., Collins, T. M., and Kerekes, J. J.,** Comparison of plankton-water chemistry relationships in three acid-stressed lakes in Nova Scotia, Canada, *Int. Rev. Gesamten Hydrobiol.,* 69, 819, 1984.

47. **Geelen, J. F. M. and Leuven, R. S. E. W.,** Impact of acidification on phytoplankton and zooplankton communities, *Experientia,* 42, 486, 1986.

48. **Marmorek, D.,** Changes in the temporal behavior and size structure of plankton systems in acid lakes, in *Early Biotic Responses to Advancing Lake Acidification,* Hendrey, G. R., Ed., Butterworths, London, 1984, 23.

49. **Schindler, D. W. and Turner, M. A.,** Biological, chemical, and physical responses of lakes to experimental acidification, *Water Air Soil Pollut.,* 18, 259, 1982.

50. **Lydén, A. and Grahn, O.,** Phytoplankton species composition, biomass and production in Lake Gårdsjön — an acidified clearwater lake in SW Sweden, *Ecol. Bull.,* 37, 195, 1985.

51. **Lam, D. C. L., Bobba, A. G., Jeffries, D. S., and Kelson, J. M. R.,** Relationships of spatial gradients of primary production, buffering capacity, and hydrology in Turkey Lakes Watershed, in *Impacts of Acid Rain and Deposition on Aquatic Biological Systems,* Isom, B. G., Dennis, S. D., and Bates, J. M., Eds., American Society for Testing and Materials, Philadelphia, 1986, 42.

52. National Research Council of Canada, Acidification in the canadian aquatic environment: scientific criteria for assessing the effect of acid deposition on aquatic ecosystems, NRCC No. 18475, National Research Council of Canada, Ottawa, 1981.

53. **Lazarek, S.,** Epiphytic algae production in the acidified Lake Gårdsjön SW Sweden, *Ecol. Bull.,* 37, 213, 1985.

54. Ontario Ministry of the Environment, Studies of Lakes and Watersheds near Sudbury, Ontario: Final Limnological Report, SES 009/82, Ontario Ministry of the Environment, 1982.

55. **DeCosta, J. and Preston, C.,** The phytoplankton productivity of an acidic lake, *Hydrobiologia,* 70, 39, 1980.

56. **Dickson, W.,** Some effects of acidification on Swedish lakes, *Verh. Int. Ver. Limnol.,* 20, 851, 1978.

57. **Pillsbury, R. W. and Kingston, J. C.,** The pH-independent effect of aluminum on cultures of phytoplankton from a Wisconsin lake, *Hydrobiologia,* submitted.

58. **Jansson, M., Olsson, H., and Broberg, O.,** Characterization of acid phosphatase in acidified Lake Gardsjon, Sweden, *Arch. Hydrobiol.,* 92, 377, 1981.

59. **Jansson, M.,** Induction of high phosphatase activity by aluminum in acid lakes, *Arch Hydrobiol.,* 93, 32, 1981.

60. **Galloway, J. N., Likens, G. E., and Edgerton, E. S.,** Acid precipitation in the northeastern United States: pH and acidity, *Science,* 194, 722, 1976.

61. **Turner, M. A., Jackson, M. B., Findlay, D. L., Graham, R. W., DeBruyn, E. R., and Vandermeer, E. M.,** Early responses of periphyton to experimental acidification, *Can. J. Fish. Aquat. Sci.,* 44 (Suppl. 1), 135, 1987.

62. **Schindler, D. W., Fee, E. J., and Ruszczynski, T.,** Phosphorus input and its consequences for phytoplankton standing crop and production in the Experimental Lakes Area, northwestern Ontario, *J. Fish. Res. Board Can.,* 30, 1511, 1978.

63. **Schindler, D. W., Wagemann, R., Cook, R., Ruszyczynski, T., and Prokopowich, J.,** Experimental acidification of Lake 223, Experimental Lake Area. I. Background data and the first three years of acidification, *Can. J. Fish. Aquat. Sci.,* 37, 342, 1980.

64. **Williams, T. G. and Turpin, D. H.,** Photosynthetic kinetics determine the outcome of competition for dissolved inorganic carbon by freshwater microalgae: implications for acidified lakes, *Oecologia,* in press.

65. **DeCosta, J. and Janicki, A.,** The population dynamics and age structure of *Bosmina longirostris* in an acid water impoundment, *Verh. Int. Ver. Limnol.,* 20, 2479, 1978.

66. **Yan, N. D. and Strus, R.,** Crustacean zooplankton communities of acidic, metal-contaminated lakes near Sudbury, Ontario, *Can. J. Fish. Aquat. Sci.,* 37, 2282, 1980.

67. **Porter, K. G.,** The plant-animal interface in freshwater ecosystems, *Am. Sci.,* 65, 159, 1977.

68. **Havens, K. and DeCosta, J.,** The effect of acidification on the biomass and population size structure of *Bosmina longirostris, Hydrobiologia,* 122, 153, 1985.

69. **Schindler, D. W.,** Experimental acidification of a whole lake: a test of the oligotrophication hypothesis, in *Ecological Impact of Acid Precipitation,* Drabløs, D. and Tollan, A., Eds., SNSF Project, Oslo, 1980, 370.

70. **Kramer, J. R.,** Acid precipitation, in *Sulfur in the Environment,* Nriagu, J. O., Ed., John Wiley & Sons, New York, 1978, 325.

71. **Stokes, P. M.,** pH related changes in attached algal communities of softwater lakes, in *Early Biotic Responses to Advancing Lake Acidification,* Hendrey, G. R., Ed., Butterworths, London, 1984, 43.

72. **Parent, L., Allard, M., Planas, D., and Moreau, G.,** The effects of short-term and continuous experimental acidification on biomass and productivity of running water periphytic algae, in *Impacts of Acid Rain and Deposition on Aquatic Biological Systems,* Isom, B. G., Dennis, S. D., and Bates, J. M., Eds., American Society for Testing and Materials, Philadelphia, 1986, 28.

73. **Müller, P.,** Effects of artificial acidification on the growth of periphyton, *Can. J. Fish. Aquat. Sci.,* 37, 355, 1980.

74. **Burton, T. M., Stanford, R. M., and Allan, J. W.,** The effects of acidification on stream ecosystems, in *Acid Precipitation — Effects on Ecological Systems,* D'Itri, F. M., Ed., Ann Arbor Science, Ann Arbor, MI, 1982, 209.

75. **Hendrey, G. and Vertucci, F.,** Benthic plant communities in acidic Lake Colden, New York: *Sphagnum* and the algal mat, in *Ecological Impact of Acidification,* Drablos, D. and Tollan, A., Eds., SNSF Project, Oslo, 1980, 314.

76. **Jackson, M.,** Filamentous algae in Ontario softwater lakes, Int. Symp. on Acidic Precipitation, Muskoka, Ontario, September 15 to 20, 1985, 117.

77. **Stokes, P. M.,** Benthic algal communities in acidic lakes, in *Effects of Acidic Precipitation on Benthos,* Singer, R., Ed., Canterbury Press, New York, 1981, 119.

78. **Lazarek, S.,** Structure and function of a cyanophytan mat community in an acidified lake, *Can. J. Bot.,* 60, 2235, 1982.

79. **Turner, P. A. E. and Stokes, P. M.,** The growth of filamentous algae *(Mougeotia* spp.) isolated from acidic lakes grown over a pH gradient from 4.5 to 6.9, Int. Symp. Acidic Precipitation, Muskoka, Ontario, September 15 to 20, 1985, 120.

80. **Stevenson, R. J., Singer, R., Roberts, D. A., and Boylen, C. W.,** Patterns of epipelic algal abundance with depth, trophic status, and acidity in poorly buffered New Hampshire lakes, *Can. J. Fish. Aquat. Sci.,* 42, 1501, 1985.

81. **Coesel, P. F. M., Kwakkestein, R., and Verschoor, A.,** Oligotrophication and eutrophication tendencies in some Dutch moorland pools, as reflected by their desmid flora, *Hydrobiologia,* 61, 21, 1978.

82. **Battarbee, R. W.,** Diatom analysis, in *Handbook of Holocene Paleoaecology and Palaeohydrology,* Berglund, B. E., Ed., John Wiley & Sons, New York, 1986, 527.

83. **Battarbee, R. W.,** Diatom analysis and the acidification of lakes, *Phil. Trans. R. Soc. London, Ser. B,* 305, 451, 1984.

84. **Charles, D. F. and Norton, S. A.,** Paleolimnological evidence for trends in atmospheric deposition of acids and metals, in *Acid Deposition: Long Term Trends,* National Academy Press, Washington, D.C., 1986, 335.

85. **Smol, J. P., Battarbee, R. W., Davis, R. B., and Meriläinen, J.,** Eds., *Diatoms and Lake Acidity,* Dr. W. Junk, Dordrecht, Netherlands, 1986.

86. **Davis, R. B.,** Paleolimnological diatom studies of acidification of lakes by acid rain: an application of quaternary science, *Quat. Sci. Rev.,* 6, 147, 1987.

87. **Charles, D. F.,** Relationships between surface sediment diatom assemblages and lake water characteristics in Adirondack lakes, *Ecology,* 66, 994, 1985.

88. **Flower, R. J.,** The relationship between surface sediment diatom assemblages and pH in 33 Galloway lakes: some regression models for reconstructing pH and their application to sediment cores, *Hydrobiologia,* 143, 93, 1986.

89. **Dixit, S. S., Dixit, A. S., and Evans, R. D.,** Sedimentary diatom assemblages and their utility in computing diatom-inferred pH in Sudbury Ontario lakes, *Hydrobiologia,* in press.

90. **Anderson, D. S., Davis, R. B., and Berge, F.,** Relationships between diatom assemblages in lake surface-sediments and limnological characteristics in southern Norway, in *Diatoms and Lake Acidity,* Smol, J. P., Battarbee, R. W., Davis, R. B., and Meriläinen, J., Eds., Dr. W. Junk, Dordrecht, Netherlands, 1986, 97.

91. **Battarbee, R. W., Smol, J. P., and Meriläinen, J.,** Diatoms as indicators of pH: an historical review, in *Diatoms and Lake Acidity,* Smol, J. P., Battarbee, R. W., Davis, R. B., and Meriläinen, J., Eds., Dr. W. Junk, Dordrecht, Netherlands, 1986, 5.

92. **Hustedt, F.,** Systematische und ökologische Untersuchungen uber die Diatomeen-Flora von Java, Bali und Sumatra, *Arch. Hydrobiol.,* Suppl. 16, 1, 1939.

93. **van Dam, H., Suurmond, G., and Braak, C. J. F.,** Impact of acidification on diatoms and chemistry of Dutch moorland pools, *Hydrobiologia,* 83, 425, 1981.

94. **Renberg, I. and Hellberg, T.,** The pH history of lakes in southwestern Sweden, as calculated from the subfossil diatom flora of the sediments, *Ambio,* 11, 30, 1982.

95. **Dixit, S. S.,** Diatom-inferred pH calibration of lakes near Wawa, Ontario, *Can. J. Bot.,* 64, 1129, 1986.

96. **Meriläinen, J.,** The diatom flora and hydrogen ion concentration of water, *Ann. Bot. Fennici,* 4, 51, 1967.

97. **Davis, R. B. and Berge, F.,** Atmospheric deposition in Norway during the last 300 years as recorded in SNSF lake sediment. II. Diatom stratigraphy and inferred pH, in *Ecological Impact of Acid Precipitation,* Drabløs, D. and Tollan, A., Eds., Proc. Int. Conf. Ecol. Impact Acid Precipitation, SNSF Project, Oslo, 1980, 270.

98. **Dixit, S. S. and Dickman, M. D.,** Correlation of surface sediment diatoms with the present lake water pH in 28 Algoma lakes, Ontario, Canada, *Hydrobiologia,* 131, 133, 1986.

99. **Renberg, I.,** Paleolimnological investigations in Lake Prastsjon, *Early Norland,* 9, 113, 1976.

100. **Berge, F.,** Diatoms and pH in some lakes in the Adger and Hordaland countries, Norway, SNSF Internal Rep., Norwegian Interdisciplinary Research Program, Oslo, IR 42/79, 1980.

101. **Dixit, S. S.,** Algal Microfossils and Geochemical Reconstructions of Sudbury Lakes: A Test of the Paleoindicator Potential of Diatoms and Chrysophytes, Ph.D. thesis, Queen's University, Kingston, 1986.

102. **Patrick, R.,** Factors affecting the distribution of diatoms, *Bot. Rev.,* 14, 473, 1948.

103. **Bennett, H. D.,** Algae in relation to mine water, *Castanea,* 34, 306, 1968.

104. **Brugam, R. B. and Lusk, M.,** Diatom evidence for neutralization in acid surface mine lakes, in *Diatoms and Lake Acidity,* Smol, J. P., Battarbee, R. W., Davis, R. B., and Merilainen, J., Eds., Dr. W. Junk, Dordrecht, Netherlands, 1986, 115.

105. **Davis, R. B., Anderson, D. S., and Berge, F.,** Paleolimnological evidence that lake acidification is accompanied by loss of organic matter, *Nature,* 316, 436, 1985.

106. **Patrick, R.,** Ecology of freshwater diatoms and diatom community, in *The Biology of Diatoms,* Werner, D., Ed., University of California Press, Berkeley, 1977, 284.

107. **Smol, J. P.,** Problems associated with the use of "species diversity" in paleolimnological studies, *Quat. Res.,* 15, 209, 1981.

108. **Taylor, M. C.,** Surficial Sedimentary Diatoms — Limnological Relationship in Canadian Shield Lakes, M.Sc. thesis, University of Waterloo, Waterloo, Ontario, 1986.

109. **Dixit, A. S., Dixit, S. S., and Evans, R. D.,** The relationship between diatom assemblages and lakewater pH in 35 Quebec lakes, Canada, *J. Paleolimnol.,* 1, 23, 1988.

110. **Dixit, S. S. and Evans, R. D.,** Spatial variability in sedimentary algal microfossils and its bearing on diatom-inferred pH reconstructions, *Can. J. Fish. Aquat. Sci.,* 43, 1836, 1986.

111. **Davis, R. B. and Anderson, D. S.,** Methods of pH calibration of sedimentary diatom remains for reconstructing history of pH in lakes, *Hydrobiologia,* 120, 69, 1985.

112. **Dickman, M. D., van Dam, H., van Geel, B., Klink, A. G., and van der Wijk, A.,** Acidification of a Dutch moorland pool — a paleolimnological study, *Arch. Hydrobiol,* 109, 377, 1987.

113. **Nygaard, G.,** Ancient and recent flora of diatoms and Chrysophyceae in Lake Gribsø, *Folia Limnol. Scand.,* 8, 32, 1956.

114. **Dixit, S. S.,** The Utility of Sedimentary Diatoms as a Measure of Historical Lake pH, M.Sc. thesis, Brock University, St. Catharines, Ontario, 1983.

115. **Tolonen, K. and Jaakkola, T.,** History of lake acidification and air pollution studies on sediments in south Finland, *Ann. Bot. Fennici,* 20, 57, 1983.

116. **Flower, R. J. and Battarbee, R. W.,** Diatom evidence of recent acidification of two Scottish lakes, *Nature,* 305, 130, 1983.

117. **Ter Braak, C. J. F.,** personal communication.

118. **Huttunen, P.,** personal communication.

119. **Kreis, R. G., Jr., Kingston, J. C., Camburn, K. E., and Cook, R. B.,** Diatom-pH relationships in the northern Great Lakes region for predicting past lake acidity, in *Paleoecological Investigation of Recent Acidification (PIRLA): Interim Report,* Charles, D. F. and Whitehead, D. R., Eds., Electric Power Research Institute, CA, in press.

120. **Hudon, C., Duthie, H. C., Smith, S. M., and Ditner, S. A.,** Relationship between lakewater pH and sedimentary diatoms in the Matamek watershed, northeastern Quebec, Canada, *Hydrobiologia,* 140, 49, 1986.

121. **Walker, I. R. and Paterson, C. G.,** Associations of diatoms in the surface sediments of lakes and peat pools in Atlantic Canada, *Hydrobiologia,* 134, 265, 1986.

122. **Huttunen, P. and Meriläinen, J.,** Applications of multivariate techniques to infer limnological conditions from diatom assemblages, in *Diatoms and Lake Acidity,* Smol, J. P., Battarbee, R. W., Davis, R. B., and Meriläinen, J., Eds., Dr. W. Junk, Dordrecht, Netherlands, 1986, 201.

123. **Steinberg, C., Arzet, K., and Krause-Dellin, D.,** Grewässerversavernug in des Bundesrepublik Deutschlund in Licht paläolimnolgischer studien, *Naturwissenschaften,* 71, 631, 1984.

124. **Stockner, J. G. and Benson, W. W.,** The succession of diatom assemblages in the recent sediments of Lake Washington, *Limnol. Oceanogr.,* 12, 513, 1967.

125. **Bradbury, J. P.,** Diatom stratigraphy and human settlement in Minnesota, *Geol. Soc. Spec. Pap.,* No. 171, 1975.

126. **Brugam, R. B.,** Human disturbance and the historical development of Linsley Pond, *Ecology,* 59, 19, 1978.

127. **Smol, J. P., Brown, S. R., and McNeely, R. N.,** Cultural disturbances and trophic history of small meromictic lake from central Canada, *Hydrobiologia,* 103, 125, 1983.

128. **Eakins, J. D. and Morrison, R. I.,** A new procedure for the determination of lead-210 in lake and marine sediments, *Int. J. Appl. Radiat. Isot.,* 29, 531, 1978.

129. **Appleby, P. G. and Oldfield, F.,** The calculation of ^{210}Pb dates assuming a constant rate of supply of unsupported ^{210}Pb to the sediment, *Catena,* 5, 1, 1978.

130. **Del Prete, A. and Schofield, C.,** The utility of diatom analysis of lake sediments for evaluating the precipitation effects on dilute lakes, *Arch. Hydrobiol.,* 91, 330, 1981.

131. **Charles, D. F.,** Recent pH history of Big Moose Lake (Adirondack Mountains, New York, USA) inferred from sediment diatom assemblages, *Verh. Int. Ver. Limnol.,* 22, 559, 1984.

132. **Davis, R. B., Norton, S. A., Hess, C. T., and Brakke, D. F.,** Paleolimnological reconstruction of the effects of atmospheric deposition of acids and heavy metals on the chemistry and biology of lakes in New England and Norway, *Hydrobiologia,* 103, 113, 1983.

133. **Charles, D. F., Whitehead, D. R., Anderson, D. S., Raymond, B., Camburn, K. E., Cook, R., Crisman, T. L., Davis, R. B., Ford, J., Fry, B. D., Hites, R. A., Kahl, J. S., Kingston, J. C., Kreis, R. G., Mitchell, M. J., Norton, S. A., Roll, L. A., Smol, J. P., Sweets, P. R., Uutala, A. J., White, J. R., Whiting, M., and Wise, R. J.,** The PIRLA project (Paleoecological Investigation of Recent Lake Acidification): preliminary results for the Adirondacks, New England, N. Great Lakes, and N. Florida, *Water Air Soil Pollut.,* 30, 355, 1986.

134. **Dixit, S. S., Dixit, A. S., and Evans, R. D.,** Paleolimnological evidence of recent acidification in two Sudbury (Canada) lakes, *Sci. Total Environ.,* 67, 53, 1987.

135. **Renberg, I. R. and Wallin, J. E.,** The history of acidification of Lake Gardsjon as deduced from diatoms and Sphagnum leaves in the sediment, *Ecol. Bull.,* 37, 47, 1984.

136. **Arzet, K., Krause-Dellin, D., and Steinberg, C.,** Acidification of four lakes in the Federal Republic of Germany as reflected by diatom assemblages, cladoceran remains and sediment chemistry, in *Diatoms and Lake Acidity,* Smol, J. P., Battarbee, R. W., Davis, R. B., and Meriläinen, J., Eds., Dr. W. Junk, Dordrecht, Netherlands, 1986, 227.

137. **Dickman, M., Dixit, S., Fortescue, J., Barlow, R., and Terasmae, J.,** Diatoms as indicators of rate of lake acidification, *Water Air Soil Pollut.,* 21, 375, 1984.

138. Ontario Research Foundation, *The Removal of Sulfur Gases from Smelter Fumes,* King's Printer, Toronto, 1949.

139. **Howard-White, F. B.,** *Nickel, An Historical Review,* Longmans, Toronto, 1973.

140. **Gorham, E. and Gordon, A. G.,** The influence of smelter fumes upon the chemical composition of lake waters near Sudbury, Ontario, and upon the surrounding vegetation, *Can. J. Bot.,* 38, 477, 1960.

141. **Smol, J. P.,** Chrysophycean microfossils as indicators of lakewater pH, in *Diatoms and Lake Acidity,* Smol, J. P., Battarbee, R. W., Davis, R. B., and Meriläinen, J., Eds., Dr. W. Junk, Dordrecht, Netherlands, 1986, 257.

142. **Smol, J. P.,** Fossil synuracean (Chrysophyceae) scales in lake sediments: a new group of paleoindicators, *Can. J. Bot.,* 58, 458, 1980.

143. **Munch, C. S.,** Fossil diatoms and scales of Chrysophyceae in the recent history of Hall Lake, Washington, *Freshwater Biol.,* 10, 61, 1980.

144. **Battarbee, R. W., Cronberg, G., and Lowry, S.,** Observations on the occurrence of scales and bristles of *Mallomonas* species in the micro-laminated sediments of a small lake in Finnish north Karelia, *Hydrobiologia,* 71, 225, 1980.

145. **Smol, J. P., Charles, D. F., and Whitehead, D. R.,** Mallomonadacean (Chrysophyceae) assemblages and their relationships with limnological characteristics in 38 Adirondack (N.Y.) lakes, *Can. J. Bot.,* 62, 611, 1984.

146. **Dixit, S. S., Dixit, A. S., and Evans, R. D.,** Paleolimnological study of LRTAP network lakes, Report, Department of Supply and Services, Quebec region, Canada, 1987.

147. **Christie, C. E., Smol, J. P., Huttunen, P., and Merilainen, J.,** Chrysophyte scales recorded in lake sediments from eastern Finland, *Hydrobiologia,* 161, 237, 1988.

148. **Smol, J. P., Charles, D. F., and Whitehead, D. R.,** Mallomonadacean microfossils provide evidence for recent lake acidification, *Nature,* 307, 628, 1984.

149. **Steinberg, C. and Hartmann, H.,** A biological paleoindicator for early lake acidification: Mallomonadacean (Chrysophyceae) scale abundance in sediments, *Naturwissenschaften,* 73, 137, 1986.

150. **Hartmann, H. and Steinberg, C.,** Mallomonadaceae (Chrysophyceae) scales: early biotic paleoindicators of lake acidification, *Hydrobiologia,* 143, 87, 1986.

151. **Tolonen, K., Liukkonen, M., Harjula, R., and Patila, A.,** Acidification of small lakes in Finland documented by sedimentary diatom and chrysophycean remains, in *Diatoms and Lake Acidity,* Smol, J. P., Battarbee, R. W., Davis, R. B., and Meriläinen, J., Eds., Dr. W. Junk, Dordrecht, Netherlands, 169, 1986.

152. **Dixit, S. S., Dixit, A. S., and Evans, R. D.,** Chrysophyte scales in lake sediments provide evidence of recent acidification in two Quebec (Canada) lakes, *Water Air Soil Pollut.,* 38, 97, 1988.

153. National Research Council, *Acid Deposition, Long-Term Trends,* National Academy Press, Washington, D.C., 1986.

154. **Charles, D. F., Whitehead, D. R., Engstrom, D. R., Fry, B. D., Hites, R. A., Norton, S. A., Owen, J. S., Roll, L. A., Schindler, S. C., Smol, J. P., Uutala, A. J., White, J. R., and Wise, R. J.,** Paleolimnological evidence for recent acidification of Big Moose Lake, Adirondack Mountains, N.Y. (USA), *Biogeochemistry,* 3, 267, 1987.

155. **Gibson, K. N., Smol, J. P., and Ford, J.,** Chrysophycean microfossils provide new ensight into the recent history of a naturally acidic lake (Cone Pond, New Hempshire), *Can. J. Fish. Aquat. Sci.,* 44, 1584, 1987.

156. **Dixit, S. S., Dixit, A. S., and Evans, R. D.,** Scaled chrysophytes (Chrysophyceae) as indicators of pH in Sudbury Ontario lakes, *Can. J. Fish. Aquat. Sci.,* 45, 1411, 1988.

Chapter 8

DIATOM STRATIGRAPHY IN ACID-STRESSED LAKES IN THE NETHERLANDS, CANADA, AND CHINA

Mike Dickman and Salem S. Rao

TABLE OF CONTENTS

I. INTRODUCTION

The preceding chapter by Dixit and Smol provides an excellent review of algal communities in acid-stressed lakes with special reference to Canadian lakes, diatoms, and chrysophytes. The present chapter reviews and emphasizes the literature dealing with diatom stratigraphy in three regions of the world experiencing heavy sulfur loading: (1) The Netherlands, (2) central Canada, and (3) east central China. It compares the rate of acidification as inferred from downcore diatom stratigraphies in each of these regions.

These three locations were chosen because they represent (1) high concentrations of SO_2 associated with the long-range transport of acidic precipitation (The Netherlands), (2) the largest point source of SO_2 emissions in the world (Sudbury, Canada), and (3) the highest concentration of small-scale, sulfur-rich coal-burning operations (Pacific Basin of China).

A. Acidification of European Moorland Pools

The Netherlands are in the very center of the acid rain region of northwestern Europe.[1] The report of forest damage by the German Federal Department of Food, Agriculture and Forestry[2] states that ''Results show that central Europe's forests are particularly badly damaged by air pollution. Of the ten countries which have provided data, five have registered (in 1986) damage to their forests of between 40 and 60%.'' The five countries reporting damage to their natural forest are The Netherlands, West Germany, Switzerland, Czechoslovakia, and Austria.

Moorland pools are the most common water bodies encountered in western Europe and are a logical choice for monitoring the impact of acid rain on aquatic ecosystems in this region. The distribution of approximately 3500 European moorland pools is associated with the presence of Pleistocene coversands. These pools are located in France, Belgium, Germany, and The Netherlands in a geological area that is highly resistant to chemical weathering. As a result the pools have low buffering capacity, low alkalinity, and low concentrations of electrolytes.[3] The majority of these moorland pools are located in The Netherlands. Achterste Goorven was chosen by a team of researchers from the Nature Management Institute in Leersum, Netherlands, as representative of many of the acid-stressed moorland pools of west central Europe.

B. Acidification of Sudbury Lakes

In Canada, the area with the lowest rainfall pH is located near Sudbury, Ontario, where metal smelting has been practiced since the latter part of the 18th century.[4] Sudbury, Ontario (area population about 150,000 people) produces approximately one fourth of the nickel ore of the world. The city of Sudbury (300 km^2 in area) lies about 400 km northwest of Toronto.

In 1883, workmen on the Canadian Pacific Railway construction crews discovered the richest copper-nickel deposits of the world near Sudbury. Nickel and copper mining began there in 1886. A new process for separating copper and nickel was adopted in 1892, touching off a major mining development. The two largest mines and smelters, Falconbridge Nickel Mines Ltd. and International Nickel Company of Canada (INCO), were founded in the early 1900s. The first observations of acid-stressed lakes in Canada were made in the Sudbury area by Gorham and Gordon in 1960.[5]

In the early 1970s, the government of Ontario ordered the Sudbury area mining and smelting companies to reduce the amounts of harmful gases which they discharged. In response, INCO built a 381-m high smokestack (the tallest in the world) which spread the gases which they discharged over a large area thereby reducing the concentration of SO_2 pollution impinging on Sudbury.

In 1986, 150 Canadian shield lakes were cored near Sudbury by Ontario Geological Survey.[6] Two of these were selected for fine sectioning (2-mm intervals) because their sediment-water interface was relatively undisturbed. The rate of lake acidification was estimated in each of these two lakes from the diatoms in their core. The methods for core diatom preparation are essentially identical to those described by Battarbee[7] as discussed by Dixit and Smol (see Chapter 7). However, the use of a 7-cm (internal diameter) Kajak corer[8] and a special hydraulic core extrusion system[9] made it possible to fine section the top 2 cm of each of these two cores at 2-mm depth intervals. Fine sectioning permitted us an opportunity to estimate very recent changes in the rate of acidification of 2 of the 150 study lakes: a small remote and unperturbed humic lake (Lake 29B) and an oligotrophic clear water lake (White Pine Lake).

C. Pacific Basin Lakes of China

In China, the area with the highest concentration of people and the highest coal-burning densities is the Pacific Basin Region.[10] Acid rain events in this area have been only rarely observed and would appear to be very localized.[10] A series of measurements during 1979 and 1980 showed that Beijing rainwater was still close to neutral. Most values were between pH 6 and 7.[11] The concentration of sulfate (SO_4) and nitrate (NO_3) was high with an average of 12 and 3.76 mg/l, respectively.[11] The investigators concluded that the high concentration of carbonate-rich suspended matter in the air neutralized the acid and kept pH high (i.e., above pH 6).

Acid rain measurements were also carried out in Shaghai and Songjiang counties starting in March of 1980. The first acid rain episodes recorded in the Shanghai area occurred on the 19th and 21st of September 1981.[10] Typically, however, the carbonate-rich loess and other carbonates in the air such as fly ash prevented the pH of the rain from dropping significantly below 6.[11]

II. METHODS

A. Bacteriological Procedures

Rao et al.[12] have indicated that sediment microbial populations and activity can be used to recognize differences between acid-stressed and unstressed lakes (see Section V.C).

Total and respiring bacteria in all water samples were estimated using the INT-formazan reduction technique.[13] A 10-ml portion of each water sample was poured into previously cleaned and sterilized test tubes and 1 ml of 0.2% aqueous INT dye (2-para-iodophenyl; 3-p-nitrophenyl; 5-phenyl tetrazolium chloride, Sigma Chemicals) was added to each of the tubes. After epifluorescence counts, respiring bacteria were counted on the same slide using a combination of UV and transmitted bright light illumination.[13] Details of the technique are recorded in Dickman et al.[9] Sulfur cycle and nitrogen cycle bacteria were estimated on all sediment samples using the five-tube most probable number procedure.[15] Nitrifying bacterial densities were estimated using *Nitrosomonas* broth. Alexander's media was used to test for denitrifying bacteria. Densities of sulfur-oxidizing bacteria (*Thiobacillus* sp.) populations were measured using postgate medium. Sulfate-reducing (*Desulfovibrio* sp.) bacterial populations were enumerated using Starky's medium with an anaerobic incubation.[12] For enumeration of bacteria that reduce organic sulfur to sulfide, an MPN medium described by Gunkel and Openheimer was used with anaerobic incubation.[12] All MPN tubes were incubated for 21 d at 20°C.

B. Sediment Coring and Water Chemistry Analysis

Sediment cores were collected near the deepest point in each of the study lakes using either a modified K. B. Gravity Corer (4.5-cm internal diameter) or a modified Kajak corer (7-cm internal diameter). To the best of our knowledge, this is the first time that large (70-mm diameter) lake sediment cores have been sectioned at 2-mm intervals in a specific attempt to analyze for downcore changes in diatom-inferred pH.

Dissolved oxygen, specific conductivity, pH, temperature, and light attenuation were measured at 1-m intervals near the center of the Canadian lakes using a Hydrolab Model 4,000 and a LiCor underwater photometer.[16]

C. Pollen Counts

Part of the thin-sectioned portions of each sediment core from each of the Canadian study lakes was placed in a beaker and stirred until the sediment was homogenous. Next, the sample mixture was dehydrated with 3 ml glacial acetic acid, centrifuged, and decanted without washing the sediment. An acetolysis solution was prepared by adding one part concentrated sulfuric acid to nine parts acetic anhydride. Of this solution, 5 ml was then added to a beaker containing the above sediment mixture and boiled for 5 min. After cooling, the mixture was centrifuged, decanted, and washed with glacial acetic acid. The final mixture was twice washed with distilled water and decanted each time. The resulting sediment was mounted on glass slides using corn syrup so the pollen grains could be rotated beneath the coverslip if required for identification purposes. A total of 400 pollen grains were counted and the number of *Ambrosia* pollen grains was expressed as a percentage of this total.

D. Lead-210 Analysis

The downcore lead-210 activity for each of the study lake sediment cores was determined with one exception (Xuanwu Lake, China). Lead-210 activity was based on the mass for each of 15 core sections analyzed from each of the lake cores. The total lead-210 activity for each core section and the estimate of unsupported activity were both based on the following procedures.

^{210}Pb was measured through its α-emitting granddaughter ^{210}Po. A 10-ml amount of wet sediment was weighed and subsequently dried at 40°C for about 12 h. Weight loss was determined and the ratio between wet weight and dry weight established. The sediment-derived material was then distilled according to the methods of Eakins and Morrison.[17] Chemical yields from this procedure varied between 95 and 99%. Additional information on lead-210 dating for sediments of clear water lakes impacted by bioturbation is provided in Section V.A.

E. Diatom Counts

Diatom enumeration was based on a minimum count of 600 diatom frustules per slide. Procedures for cleaning, mounting, and counting the diatoms were described by Dickman et al.[18] References used in the identification of the diatoms and the assignment of pH indicator status included Beaver,[19] Camburn and Kingston,[20] Charles,[21] Cleve-Euler,[22] Foged,[23] Hustedt,[24,25] Patrick and Reimer,[26,27] and Germain.[28]

The precision of the diatom-inferred pH technique was estimated by making five replicate slides of the sediment diatoms for each of the study lakes at one to five randomly selected sediment core depths. All slides were coded to avoid unconscious bias and 600 frustules were counted per slide. The results of these coded slide replicate tests were used to calculate the range about the mean. This is represented by a vertical line atop the histogram downcore profiles in the figures accompanying this chapter.

F. Diatom-Inferred pH

The downcore pH for each lake was inferred from the relative percentages of acid to alkaline diatoms at each sediment depth. This ratio was modified by Renberg and Hellberg[29] and Dickman et al.[18] and referred to as Index B.

$$\text{Index B} = \frac{(\% \text{ circumneutral diatoms}) + (5 \times \% \text{ acidophilic diatoms}) + (40 \times \% \text{ acidobiontic diatoms})}{(\% \text{ circumneutral diatoms}) + (3.5 \times \% \text{ alkaliphilic diatoms}) + (108 \times \% \text{ alkalibiontic diatoms})}$$

Although it is frequently difficult to distinguish between circumneutral and pH-indifferent diatoms, the latter contribute little to the inference of lake pH and should be ignored as they increase the variance without increasing the precision of the diatom-inferred pH estimates.[18]

III. RESULTS AND DISCUSSION

A. The Ideal Lake for Stratigraphic Analysis

The ideal lake for constructing a diatom stratigraphy is one with anaerobic, varved sediments in which sediment resuspension and bioturbation are negligible or nonexistent.[30] Because unbioperturbed lakes are rare, it must be determined whether there is any value in coring a lake with sediments in which considerable mixing or bioturbation has occurred (see Section V).

One of the first paleolimnologists to study downcore changes in the abundance of diatoms in shallow lakes where the sediments were both bioperturbed and physically mixed was Moss.[31] His classic study of the paleoecology of a shallow broad near Norfolk, England, established the fact that the sediments of shallow water habitats could be used for "coarse-grained" (decade by decade) analyses of diatom stratigraphy.[31,32] Following this study, a team of paleolimnologists in The Netherlands selected a number of similarly perturbed moorland pools to determine the rate of stratigraphic change in their downcore diatom profiles.

B. Diatom Stratigraphy of a Moorland Pool and Inferences of the Rate of Acidification

The moorland pools of Belgium and west central Europe are severely impacted by acid rain.[9,33-36] Moorland pools are characteristically oligotrophic and are located in regions with sandy soils.[34]

During medieval times these lands were heavily logged. After deforestation their soils supported a vegetation dominated by the heath plants *Calluna vulgaris* and *Erica tetralix*.[37] Sheep overgrazing in these areas resulted in the depletion of soil nutrients resulting in the gradual development of aeolic drift sands which were colonized by a variety of grasses.[37] In the late 17th century, sheep farming decreased and these lands were planted with Scots pine (*Pinus sylvestris*).[38]

The moorland pool selected for study (Achterste Goorven) displayed a perched water table, was located in the area of unconsolidated sands and podzols, and was fairly typical of many west central European moorland pools. Achterste Goorven had been previously

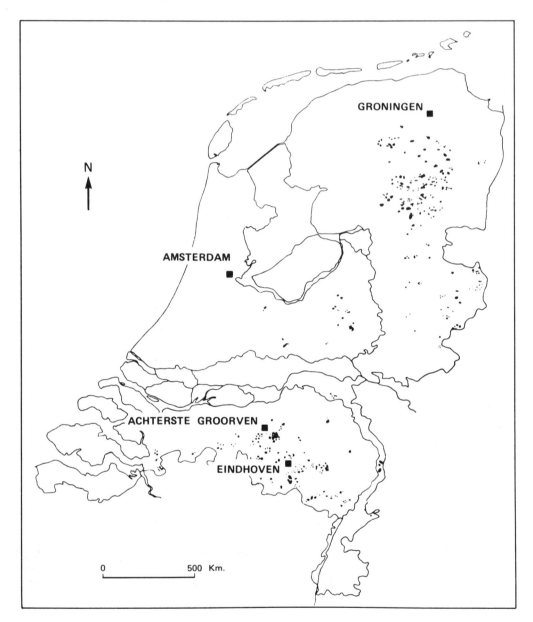

FIGURE 1.   Location map for the moorland pond Achterste Goorven (The Netherlands). Dots represent the location of other moorland ponds.

studied by Heimans[39] from 1909 to 1947. Heimans' pH analyses of its waters indicated that it had reached its highest pH in the first half of this century[34] and thereafter began to decline in pH.[9,34]

1. Site Description

Achterste Goorven is located near Oisterwijk in the southeastern portion of The Netherlands (Figure 1). It has a surface area of 2.3 ha and a mean depth of 0.6 m. The pool has negligible contact with groundwater as indicated by its low electrolyte content (35 to 68 μS cm^{-1}) and low calcium content (1 to 11 mg l^{-1}). During the study period (1979 to 1984) its pH ranged from 3.4 to 4.5 with a mean of 3.8 and its sulfate concentrations ranged from 12 and 58 mg l^{-1} with a mean of 35 mg l^{-1}.[9]

The acidobiontic diatom, *Eunotia exiqua,* was the most numerically important diatom in the surficial sediments of Achterste Goorven.[9] The dominance of *E. exiqua* in acidified moorland pools[34] is matched only by a few lakes in other parts of Europe.[29] The other abundant diatom taxa in the 0- to 1-cm sediment depth sample from Achterste Goorven were very similar to those described for acidified lakes in Scandinavia and Scotland (e.g., *E. incisa, Tabellaria quadriseptata,* and *Navicula hoefleri.*)[39]

The rate of acidification in Achterste Goorven was among the highest reported in the published literature. Both sediment diatom-inferred pH and direct observations of pH demonstrated a decline in pH from 5.8 to 3.8 over the last 60 years (Figure 2).

Two major factors were responsible for this high rate of acidification. First, in dry years, large parts of the bottom of this shallow pool dried up and subsequently the reduced sulfur compounds in the sediments were oxidized to sulfuric acid. This acidification process was reflected by a noticeable drop in diatom-inferred pH as the moorland pool refilled. Similarly, Renberg[40] reported that sulfides in the marine sediments of Lake Blamissusjon were oxidized to sulfur, sulfates, and ultimately to sulfuric acid following their exposure to the air. Second, Belgium and The Netherlands are located in the very center of the European acid rain region.[41] Measurements of the pH in the rainwater at the nearest precipitation monitoring station, located approximately 20 km from Achterste Goorven, gave a mean pH of precipitation for the period 1978 to 1982 of 4.4.[42] The mean SO^{-3} S for the same period was 7.1 mg l^{-1}. This resulted in a very high deposition rate of about 160 kg S ha^{-1} $year^{-1}$.[42] Accordingly, the mean sulfate concentration in the pool during the period 1979 to 1984 was very high with 12 mg l^{-1} as an average of 23 measurements.[9]

Battarbee,[7] reviewing the available studies on diatoms and lake acidification, concluded that the published evidence generally indicated that lake acidification was due to an increase in the acidity of precipitation resulting from emissions from fossil fuel combustion. The average pH decline during the last century as inferred from diatom data from 34 lakes in Scandinavia and North America was 0.6.[7] The onset of acidification began as early as 1850.[7] Achterste Goorven began acidifying after World War I and during the intervening years it has declined in pH until the mid-1970s from an initial pH of 5.8 to a minimal pH of 3.8.

2. Conclusions from Moorland Pool Studies

Berge,[43,44] Renberg and Hellberg,[29] Tolonen and Jaakkola,[45] Davis et al.,[46] Flower and Battarbee,[47] and Battarbee et al.[48] investigated the changes in species composition of diatom assemblages in cores taken from acidifying lakes in Scandinavia and Scotland. In all such cases, acidophilous and acidobiontic species such as *Anomoeoneis serians, E. bactriana, E. denticulata, N. subtilissima,* and *Semiorbis hemicyclus* were increasing from the bottom to the top of the sediments. *T. binalis* increased with acidification both in Scandinavia and Scotland, while *T. quadriseptata, N. hoefleri,* and *E. incisa* seem to be more dominant in the surficial sediments of the Scottish lochs. *E. exiqua* generally was of minor importance in the above area.

The changes in pH recorded by Heimans[39] during his studies of Achterste Goorven from 1909 to 1947 were consistent with its downcore diatom-inferred pH profile (Figure 2). Although downcore ''slurring'' (Section V.A) did occur in Achterste Goorven, the overall sediment core pattern was consistent with the historical record as determined by Heimans.

This study also supported the contention of Moss[32] that valuable coarse-grained stratigraphic information can be gleaned from the sediments of shallow water bodies.

C. Diatom Stratigraphy of Two Sudbury Lakes

A long-term historical record of pH is the exception rather than the rule. In the Canadian and Chinese acid rain study areas which are described in this chapter, the date on which the study lakes first began to acidify was not known. The extent to which they acidified

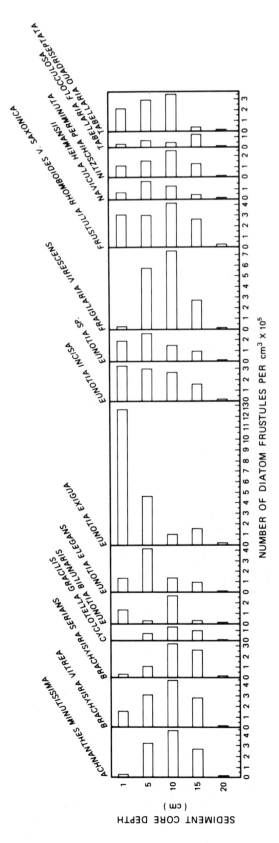

FIGURE 2.   Downcore changes in the relative abundance of the 15 most common diatoms from the sediments (0.20 cm) of Achterste Goorven.

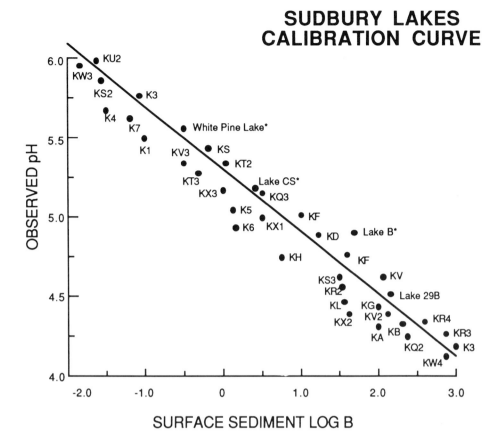

FIGURE 3. Regression curves for Sudbury lakes for the relationship between observed and diatom-inferred pH for 36 study lakes in the Sudbury region and 2 in the Algoma region (Lake B and Lake CS).

and the rate at which they acidified was also not known. For these reasons downcore diatom stratigraphic analyses were undertaken in order to answer these questions.

1. Sudbury Lake Downcore Diatom-Inferred pH Calibration Curves

In order to use diatoms as indicators of past lake pH, it is necessary to establish how recently deposited fossil diatom assemblages vary with lake pH.[49] To this end, 150 lakes in the Sudbury area were cored and their pH measured. The relationship between diatom-inferred pH and observed pH was determined (Figure 3). This relationship was sufficiently similar to Renberg and Hellberg's[29] Log B that the latter index was adopted here to permit downcore diatom pH comparisons between lakes in each of the three parts of the world which were examined in this study.

In Sudbury, a humic and a clear water lake (Figure 4) were chosen for fine-scale (2-mm) examination. These 2 lakes were selected from the 150 Sudbury study lakes because the sediments of these 2 lakes were essentially undisturbed by bioperturbation or physical mixing. As a result, high resolution temporal variation in diatom-inferred pH was studied in these two Canadian study lakes.

The top 4 mm of sediment in Lake 29B represents the total accumulation of diatoms during the last 6 to 10 years. The significant ($p < 0.01$) reduction in the abundance of all but two pH indicator diatoms found in the top 4 mm of sediment of Lake 29B was interpreted as indicative of a trend toward lake *deacidification*. To the best of our knowledge, this is one of the few times that diatoms have been used in North America to provide evidence of lake deacidification.

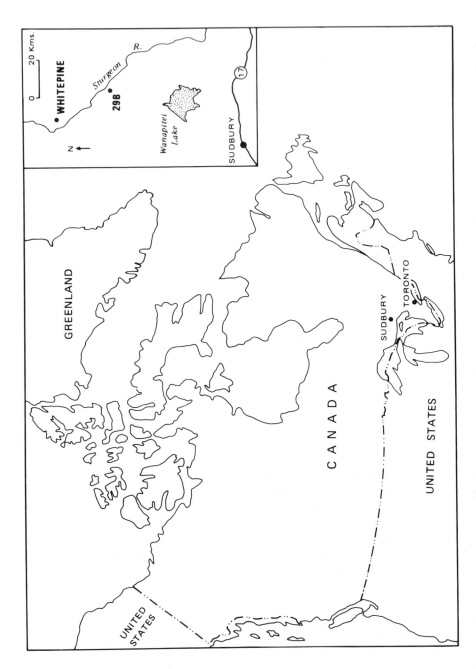

FIGURE 4. Location map for White Pine Lake and Lake 29B near Sudbury, Ontario (inset) plus a map to represent the general location of the Sudbury study area in Canada.

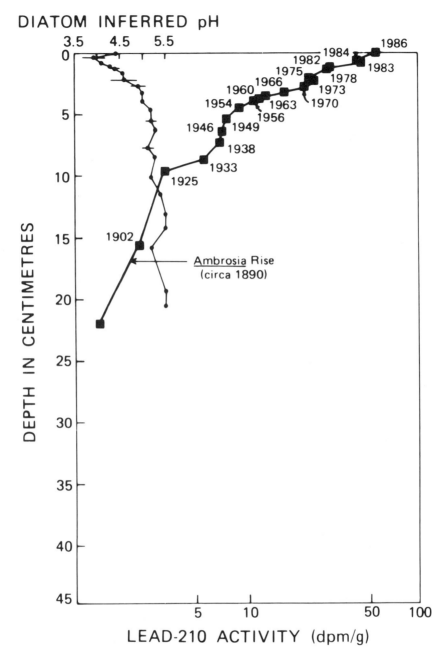

FIGURE 5.   Lake 29B (Bog Lake, Sudbury, Ontario) *Ambrosia* rise, downcore diatom-inferred pH (horizontal lines represent one SD; n = 3), and lead-210 profile estimated dates.

2. Diatom-Inferred Evidence for Lake Deacidification in the Sudbury Region

The most interesting feature of the downcore diatom profile for Lake 29B (Figure 5) was the sharp decline in the relative abundance of seven acid-loving taxa in the top 4 mm (about 1983 to 1986). A decrease in the relative abundance of three acidobiontic diatoms, (*Actinella punctata, Fragilaria acidobiontica,* and *T. quadriseptata*) as well as four acidophilic diatoms (*E. serra, F. constricta, F. hungarica,* and *Melosira lirata*) in the top 4 mm of the Lake 29B sediment core (Figure 6) was interpreted as a strong indication of deacidification during the last 6 years. By way of comparison, the downcore diatom-inferred pH of the much larger White Pine Lake has changed very little over the last 200 years (Figure 7).

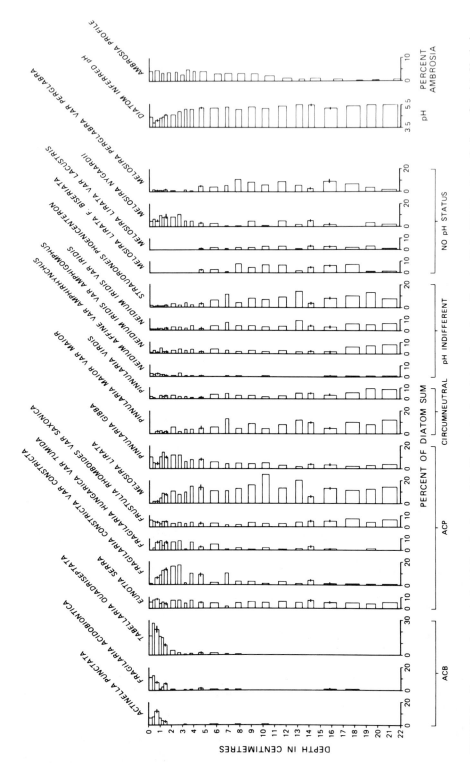

FIGURE 6. Downcore changes in the relative abundance of the 19 most abundant diatoms from the sediments of Lake 29B (Bog Lake, Sudbury, Ontario). Vertical bars represent ranges (n = 3). Diatom-inferred pH and percent *Ambrosia* pollen are provided by two histograms at the far right.

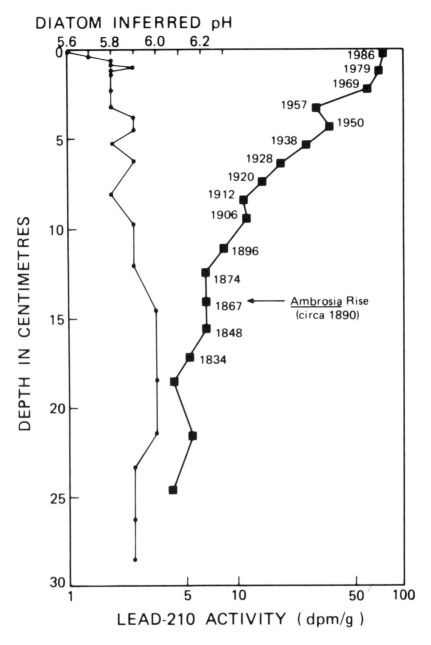

FIGURE 7.   White Pine Lake (Clear Water Lake, Sudbury, Ontario) *Ambrosia* rise, down-core diatom-inferred pH, and lead-210 profiles.

In lakes within the Sudbury basin, recent increases in pH were often attributed to reduced SO₂ emissions from the smelters.[50-52] The closing of the Coniston smelter near Sudbury in 1972 was the key factor leading to the dramatic example of lake deacidification which was attributed to lake pH recovery in one localized part of the Sudbury basin near the Coniston smelter.[52]

It is encouraging to note that the trend from 1930 to 1970 toward lower pH levels in aquatic ecosystems in the Sudbury area is reversible. This is consistent with a recent study which reported rapid recovery (deacidification) of many acid-stressed lakes in western Sweden.[53]

In 1958 (± 5 years) the diatom-inferred pH of Lake 29B was 5.2 (Figure 5). Thereafter diatom-inferred pH decreased upcore to 3.9. However, during the last 6 years, the pH of Lake 29B rose to 4.5 (Figure 5). The observed pH of Lake 29B in 1986 was 4.8. Thus, it would appear that Lake 29B is currently undergoing deacidification.[54] This is consistent with observations of Hutchinson and Havas[52] and Nriagu and Rao,[54] who noted that lakes in the Sudbury Region are receiving significantly lower levels of acid precipitation than they received 10 years ago.

3. Lake Deacidification and Sulfate Reduction

Schindler et al.[55] reported that after the addition of sulfuric acid to Lake 223 in 1976 it was rapidly reduced by sulfate reducing bacteria to sulfides. Based on our observations of the increase of *Desulphovibrio* spp. in lake sediment cores,[9] we concluded that some of this reduced sulfur combines with ferrous ions in the anaerobic portion of the lake hypolimnion to form FeS which increases the anaerobic hypolimnetic alkalinity of the lake (see also Section V.C).

Cook and Schindler[56] stated that extensive bacterial sulfate reduction and isotopic fractionation occurs in anoxic zones of lakes with sulfate concentrations as low as 30 to 40 mmol l^{-1} (3 to 4 ppm of $SO_4=$). This sulfate reduction contributes up to 70% of the total reduced sulfur to their sediments.[56] However, below levels of 20 to 30 mmol l^{-1}, sulfate reduction may no longer be energetically favorable. These were the lowest concentrations of sulfate found in the bottom sediment pore waters of Lake 223.[56]

Since sulfate reduction by pure cultures of sulfate reducers (*D. desulphuricans*) has been reported at these low concentrations,[57] it is not clear at what concentration sulfate becomes a limiting factor and at what concentration the bacteria involved in sulfate reduction reach their lower sulfate concentration limits. Our observations of upcore increases in the abundance of *D. desulphuricans* in Lake 29B indicated that the sediments with the highest sulfur content were associated with the highest densities of *Desulphovibrio*.

We were able to conclude that during periods of hypolimnetic hypoxia, the *Desulphovibrio* spp. in Lake 29B reduced sulfate to sulfide and by so doing generated considerable alkalinity. Previous studies have also reported alkalinity production in lake sediments affected by acid rain.[9,58]

4. Production of Alkalinity in Anaerobic Sediments

Ambient lake water alkalinity is not a good predictor of buffering capacity of lake sediments. The principle source of acid neutralization is from the release of calcium ions.[59] Sources of sedimentary calcium ion fluxes vary from 0 to 130 meq m^{-2} year^{-1}. These fluxes are the result of mineral weathering and advection of groundwaters. Porewater pH is confined to a narrow range close to the pK of carbonic acid (H_2CO_3) and indicative of a balance between excess CO_2 resulting from diagenetic reactions and the kinetics of silicate weathering. A significant portion of lake sediment porewater alkalinity is associated with Fe^{+2}, Mn^{+2}, and NH_4^{+1}.[60]

The primary factor determining redox status of surficial sediments is the availability of organic matter. Unproductive lakes exhibit oxidized surficial sediment potentials above 200 mV while those of more productive lakes are typically less than 200 mV.[61] As noted earlier, sulfate reduction in sediments requires organic matter and ultimately leads to alkalinity production.[58] Sulfate consumption rates range from 5 to 160 meq m^{-2} year^{-1}. Permanent alkalinity from SO_4 reduction to sulfides results from a combination of each of the factors discussed above.[59]

Brugam and Lusk[49] failed to acknowledge the importance of alkalinity generated by the reduction of sulfates to sulfides under anaerobic conditions in their study of acid mine lakes. It is our impression that this factor and not the simple "weathering out" of sulfide minerals in their acid mine lakes resulted in the observed rapid deacidification of these lakes.

The acid mine spoil lakes which they studied were divided into two distinct groups: those with a pH below 4.5 and those with a pH above 6.0. Very few lakes of intermediate pH (4.5 to 6.0) were observed due to the fact that two different buffering systems appeared to be operating in these two lake groups.[49] Thus, lakes displaying a pH below 4.5 were dominated by *Pinnularia biceps, P. terminata, P. obscurans, Navicula* sp. I (possibly *N. minutissima*), *E. exigua, Frustulia rhomboides,* and *Anomoeneis serians.* Above a pH of 6.0, *Cyclotella menenghiana, C. steligera, Mastogloia smithii, N. halophila, Gyrosigma accuminatum,* and *Rhopalodia gibba* formed the dominant diatom assemblage.[49]

5. Sulfate Reduction

Sulfate reduction occurs rapidly in anaerobic sediments when there is plenty of organic matter to drive the endothermic reduction reaction. If the quantity of organic matter is low, sulfate reduction occurs at a far lower rate. The reduction reactions are carried out by obligate anaerobic dissimilatory sulfate-reducing bacteria. The resulting free H_2S is rapidly combined with metals to form metal sulfides which (in the absence of oxygen) are extremely stable and relatively insoluble. If oxygen is added the metal sulfides are rapidly oxidized to elemental sulfur and sulfuric acid. If the sulfate released by sulfur oxidation is assimilated by algae, the amount of acid formed is generally offset by the release of hydroxyl ions or bicarbonate ions from the algae during photosynthesis. Thus, the importance of the microbial sulfur cycle in lake sediments receiving airborne sulfur pollutants relates mainly to the mobility of the sulfate ion and cation loading rates.

6. White Pine Lake, Sudbury

The pH of White Pine Lake (Figure 7) ranged between 5.7 to 5.9. Unlike Lake 29B, White Pine lake is a clear water lake (vertical extinction coefficient for White Pine Lake was 0.45 in late summer). Light at 8 m in White Pine Lake was sufficient to support the growth of aquatic moss. Dissolved oxygen was essentially saturated from top to bottom (orthograde dissolved oxygen (DO) profile *sensu* Hutchinson.[62] The specific conductivity (36-40 μS cm^{-1}) and low alkalinity (<5 mg l^{-1}) of the lake were characteristic of very oligotrophic lakes possessing minimal buffering capacity.

Lake 29B (Figures 4 and 5) was dystrophic (humic) with a clinograde DO profile (mud-water interface DO = 0.6 mg/1). Its Secchi transparency was 3 m and its vertical extinction coefficient was 1.2. For much of the year Lake 29B had no oxygen at its mud-water interface and as a result sediment bioturbation was negligible. The pH of the lake in 1986 ranged between 4.8 and 5.1.

7. White Pine Lake Downcore Chronology

Each of the two Sudbury area study lakes were cored with a modified Kayak corer[8] and a K.B. corer. Sedimentation rates were established for each of the two study lakes using their lead-210 profiles and their *Ambrosia* rise depths (Figures 5 and 7). In Lake 29B, the *Ambrosia* rise (about 1890) occurred at a sediment depth of 17 cm and the lead-210 estimate of this age also occurred at a depth of 17 cm (Figure 5).

The correspondence between the *Ambrosia* rise and the lead-210 estimate of this age was closer in Lake 29B than in the clear water, White Pine Lake. The *Ambrosia* rise (circa 1890) occurred at a depth of 14 cm in White Pine Lake and the lead-210 age estimate for 14 cm was 1867 (Figure 7). The reasons for this lack of correspondence between lead-210 and the *Ambrosia* rise estimated age in the clear water, White Pine Lake, may (in part) be due to outlet discharges of lead-210.[63] It is possible that the humic material in the brown water lakes acts to sequester the lead-210 in their sediments, reducing its loss via outlet discharges.[63] Thus, the reason for the close correspondence between the lead-210 estimates and the *Ambrosia* rise (circa 1890) in the humic lake (29B) may reflect the role of humic matter in

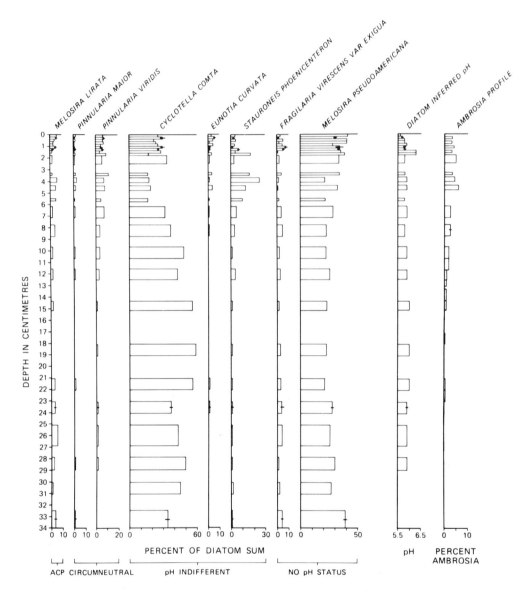

FIGURE 8.   Downcore changes in the relative abundance of the eight most abundant diatoms from the sediments of White Pine Lake (Clear Water Lake, Sudbury, Ontario). Vertical bars represent ranges (n = 3). Diatom-inferred pH and percent *Ambrosia* pollen are provided by the two histograms at the far right.

preventing lead bound organics from leaving the sediments.[63] Another important factor in this regard (see also Section V.A) was that the surface sediments of Lake 29B were essentially anaerobic and thus their sediments were undisturbed by burrowing invertebrates. This would have increased the precision of the lead-210 and *Ambrosia* pollen estimates of age in Lake 29B.[64]

The commoner diatoms in the cores from the two Sudbury study lakes (i.e., those with abundance of more than 5% of the total) were grouped into six categories and plotted (Figures 6 and 8). The diatom-inferred pH was based on the ratio of acid to alkaline diatoms as previously described. No alkalibiontic diatoms (pH range > 8.5) were observed in the Sudbury area study lakes. However, a group of circumneutral diatoms (pH range of 6 to 8) and a group of pH-indifferent diatoms (found in lakes of various pHs) were observed.

The dominant diatom in the White Pine Lake sediment core was the pH-indifferent diatom *Cyclotella comta* (Figure 8). This planktonic diatom often comprised as much as 50% of the diatoms found in the White Pine Lake sediments. It peaked in its relative abundance approximately 100 years ago (Figure 7). *Melosira pseudoamericana* which is described in the PIRLA lakes as displaying no pH status[20] was the next most abundant taxon in White Pine Lake. Neither *C. comta* nor *M. pseudoamericana* changed markedly downcore during the last 200 years. Thus, diatom-inferred paleo pH in White Pine Lake remained relatively constant over the last century (Figure 7).

By way of contrast, the paleo pH of the nearby humic lake (29B) declined sharply during the last 15 to 20 years from 5.4 ± 0.3 to 4.5 ± 0.3 (Figure 5). This was reflected by the sudden increase in the relative abundance of three acidobiontic species at 2 to 3 cm (*Actinella punctata, Fragilaria acidobiontica,* and *Tabellaria quadriseptata*). Moreover, at the same time that the acidobiontic and acidophilic taxa were increasing in their upcore relative abundance, alkaliphilic taxa and two circumneutral taxa, *Pinnularia maior* and *P. viridis,* were decreasing (Figure 6).

8. Sudbury Lakes Deacidification: Some Conclusions

The water in lakes (even those in granitic basins) does not behave like distilled water. When one stops pouring acid into a container of distilled water, its pH stabilizes and remains low and constant over time. By way of contrast, when one stops acidifying a lake, its pH does not remain low and constant but rebounds in the direction of its preacidified condition.

In lakes where negligible bioturbation occurs, it is possible to use diatom-inferred pH techniques based on thin sections (2-mm thickness) to infer recent changes in diatom stratigraphy. In the case of Lake 29B it was possible to conclude that biological processes operating in the lake basin had acted to return its pH to levels approaching those which had occurred in its preacidified state. These conclusions point to the fact that lakes which have been acidified by the addition of sulfuric acid to waters will rapidly recover once this acid influx is stopped. This underscores the need for source removal of SO_2 emissions as a critical step in reversing the trend in lake acidification.

Over the last 5 years, substantial SO_2 reductions have occurred in Ontario and Quebec. This culminated in July of 1986 in the signing of the international agreement in Helsinki to reduce sulfur dioxide emissions by an additional 30% by 1994.[65] According to the Environment Canada Annual Report for 1985-1986, ''acid rain reduction remained on top of the Department's list of priorities''.[65] It is our contention that evidence of lake deacidification for small (humic) lakes of the Sudbury area reflects the reduction of SO_2 loading levels which has occurred over the last 5 to 10 years. In Lake 29B, the first evidence of deacidification occurred at 4 mm (circa 1983). Below these depths, pH increased downcore until it plateaued at 5 cm (circa 1946). In Lake 29B, the diatom-inferred pH at 5 cm was 5.3 to 5.4 (Figure 5). This pH was significantly higher than the diatom-inferred pH at 4 mm (p <0.05). Thus, in Lake 29B, a statistically significant upcore increase in diatom-inferred pH during the last 30 years, when diatom-inferred pH fell from 5.5 to 4.8 during the period 1950 to 1980 (5.0 to 0.4 cm), was followed by a statistically significant reversal in the rate of acidification in the top 4 mm of sediment. Thereafter pH increased to its present day observed pH of 5.2 (0.4 to 0.0 cm).

While the nearby (much larger) study lake, White Pine Lake, was presumably exposed to similar SO_x and NO_x loading, there was only a slight decrease in its pH during the last 30 years and there was no evidence of deacidification. Thus, we concluded that lake volume is a major component variable which must be considered in determining the rate of lake acidification and deacidification.

These results hold out the hope that by reducing SO_2 loading we may be able to *rapidly* reverse the process of lake acidification in many of the small basin lakes which have been altered by acid rain. We feel that the reduction in SO₂ loading in the Sudbury area during the last decade was related to recent increases in diatom-inferred pH as exemplified by Lake 29B.

D. Lake Acidification in China

In the late 1970s the Chinese government started construction of a series of large mine-mouth power plants in Jilin, Liaong, Hebei, Shaanxi, and Shandong provinces.[10] The continuing serious shortages of electricity and the high costs of the long lead times for hydrostation construction led to proposals for building a truly giant concentration of coal-fired stations in Shaanxi Province, the richest coal bearing province in China. The mine-mouth power plants planned in the province will (in time) lead to a combustion of tens of millions of tons of bituminous coal per year.[14] The large sulfur dioxide emissions resulting from this would be carried in a southerly and southeasterly direction in winter and a northerly and northwesterly direction in summer, affecting, in the first case, farmlands of the North China Plain and, in the second place, Nei Monggol pastures.[14] Moreover, the valley of the Fen He in Shaanxi Province has very frequent periods of calm (over 30% of the time throughout the year and up to 70% in some locations).[66] This means that reduced atmospheric mixing and more common and stronger inversions will result in a greater likelihood of higher concentrations of gaseous emissions.

Prevailing winds from the lignite fields of Jilin, where another series of large mine-mouth stations is planned, would carry the summer emissions over the rich boreal forest area located in the Da Hinggan Region of China.[9] In addition, increased SO₂ releases from coal combustion anywhere in China would have, in time, major destabilizing effects on aquaculture, which is now again expanding.[14] It was for this reason that the following Xuanwu Lake acidification study was undertaken in Nanjing China in 1986.

Acid rain precipitation with pH well below the natural value of 5.6 is a phenomenon as old as the large-scale, concentrated combustion of coal, but only during the past 2 decades has its extent and effects been studied in detail in Europe, North America, and Japan.[14] As long as the pollutants causing the effect, oxides of sulfur above all, are released in relatively small volumes from low stacks, the problem remains just a localized nuisance. Major difficulties arise when massive releases from tall stacks of coal-fired power plants or ore smelters are carried by prevailing winds, often for hundreds of kilometers, to affect large areas downwind from the source. As a result, previously unpolluted regions far from any major industries may turn out to have greater frequency of a more acid rain than the locations surrounding the pollution sources. Illustrations of this occurrence can be found in southern Norway and Sweden, where the acid rain is caused mostly by the emissions from England which are carried across the North Sea, or in New England and the Canadian maritimes, which are affected by pollutant releases from the industrial heartland of the U.S. and Canada.[14]

1. Yangtze River Flood Plain Lakes

In 1960, a symposium was convened by the National Geographic Society to discuss the results of limnological studies carried out in the southern part of Jiangsu Province.[64] It was concluded that the lakes of this area are generally characterized by slightly elevated pH (7.2 to 8.4), moderate specific conductivity (100 to 300 μS cm^{-1}), and high productivity resulting from high algal standing crop associated with an influx of nutrient-rich runoff. The data on most of these lakes is published in Chinese (e.g., Bulletin of the Nanjing Institute of Geography 1981 to 1986) and apart from the excellent papers of Shi and Laing,[67] Wong et al.,[69] and Chang[70] very little has been published in international journals concerning the lakes of the eastern central China plain. Herdendorf[71] came to the same conclusion after his review of the English language literature on the lakes of China.

2. Xuanwu Lake

Since very little has been published in international journals on the nature of the smaller lakes of China,[72] a team of scientists lead by Xu of the Nanjing Normal University was assembled to study Xuanwu Lake. Xuanwu Lake is small (3.7 km²) and shallow (1.4 m mean depth and 2.3 m maximum depth). The lake is located in the northern portion of the City of Nanjing, the ninth largest city in China (population 3 million). The lake was chosen for paleolimnological investigation and acid rain study because of its long recorded history and because it had recently (1952) been dredged down to its hard pan clay base making it possible to date the material at the base of its sediment core. The sediment core from Xuanwu Lake was taken from near the deepest point in the lake (2.1 m; station 15, Figure 9).

Information concerning the history (both long term and recent) for this was easily acquired because the lake has played an important role in the development of the City of Nanjing. Xuanwu Lake was used by Liu Son Wu Di during the Liu Song Dynasty (420 to 479) of the Southern Dynasty Period (420 to 589) to train his navy.[74] In 463 he held an extravagant naval review on Xuanwu Lake. In the early Ming Dynasty (1356 to 1398), buildings were erected on the shores of Xuanwu Lake for keeping and maintaining the national census and taxation records.[74] Few lakes have such well-documented histories stretching back over $1\frac{1}{2}$ millennium.

3. Diatom Stratigraphy of Xuanwu Lake

Xuanwu Lake was cored on May 28, 1987 with a K.B. gravity corer (4.5-cm internal diameter). The resulting core was sectioned at 1-cm intervals over its 16-cm length in a fashion similar to that described in Dickmen et al.[9] At each sectioned depth interval (1 cm), 600 diatoms were counted. When broken frustules were encountered, they were included in the diatom count if the frustule was identifiable and if it constituted more than half of a complete valve.

Pinnularia viridis, Amphora ovalis, and *Synedra vaucheriae* were present at moderate densities at the base of the core from Xuanwu Lake (12 to 13 cm). These same diatoms were virtually absent near the surface of this core (Figure 10). Near the top of the core (0 to 3 cm), pollution-tolerant taxa such as *Nitzchia palea, N. intermedia,* and *Melosira granulata v. angustissima f. curvata* became abundant. Progressive eutrophication leading to the increase in the relative abundance of numerous eutrophic indicator species was evident (Figure 10).

The top 13 cm of the 16-cm long core from Xuanwu Lake was composed of an algal-rich material which was intermixed with fine silts and clays. The bottom 3 cm of our core penetrated the hard pan clays of the lake which we believe were exposed during the dredging of the lake during the early 1950s as previously mentioned. The average sedimentation rate during the last 35 years in Xuanwu Lake was estimated on this basis as 0.37 cm year^{-1}.

A common method of determining dates of diatom-inferred pH shifts where cesium and lead isotope data are unavailable is to find the deepest layer in the core which contains diatoms and to assume that it was deposited at the time of lake formation.[25] In Xuanwu Lake, the basal date (1952) was ascribed to the presence of hard pan sediments resulting from lake dredging.[73]

The major downcore diatom-inferred pH changes which occurred during the last 35 years are reported in Figure 10. In brief, the increase in hypereutrophic indicator species at the top of the core was interpreted as indicative of a trend toward progressive eutrophication. There was no evidence of lake acidification during the last 34 years (1952 to 1986). As *N. palea, N. intermedia, N. apiculata,* and *M. granulata* sp. increased upcore, benthic diatoms such as *Amphora ovalis* and *Surirella robusta,* became rarer. The virtual disappearance of the benthic species in the top 2 cm of the core was probably associated with the reduction of the photic zone thickness as the lake became more eutrophic and consequently more

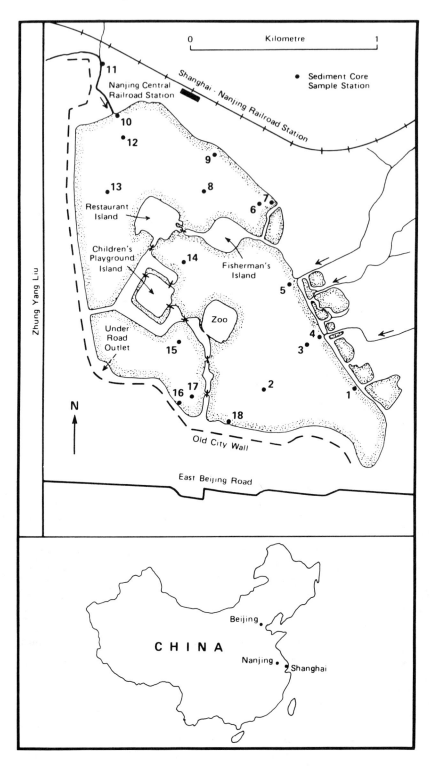

FIGURE 9.   Nanjing, China location map with inset of the morphometry and core site of Xuanwu Lake in Nanjing. The sediment core was taken at station 15.

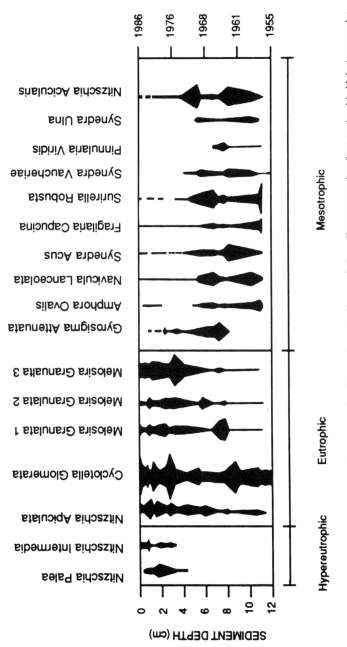

FIGURE 10. Downcore changes in the 17 most abundant diatoms from the Xuanwu Lake sediment core taken from station 14. *Melosira granulata* 1 refers to *M. granulata* of the nominate variety. *M. granulata* 2 refers to *M. granulata* var. *angustissima*, and *M. granulata* 3 refers to *M. granulata* var. *angustissima* forma *curvata*.

densely colonized by phytoplankton resulting in reduced light penetration to the bottom of the lake.

The diatom assemblage in the top layer of sediment of Xuanwu Lake (0 to 1 cm) was very similar to the assemblage described by Seenayya[75] for the eutrophic Hydrilla Pond which he studied in India. Seenayya identified 50 species of diatoms, many of which were eutrophic indicator species similar to those which we identified in the uppermost sediment layers of Xuanwu Lake.

The overall pattern of upcore diatom succession in Xuanwu Lake was consistent with the hypothesis that the lake was undergoing progressive eutrophication. This idea was also supported by Liu,[76] who carried out phytoplankton counts for Xuanwu Lake for the period 1976 to 1986. There was no evidence that Xuanwu Lake was acidifying.

4. Lake Acidification and Lake Productivity Relationships

Studies of the relationship between pH and lake productivity are too numerous to detail here. One of the most thorough of these was conducted by Berzins and Pejler[77] in Sweden from 1945 to 1982. The authors concluded that eutrophic lakes rarely if ever had a pH below 7.0. Furthermore, they were colonized by organisms with a pH optima at or above pH 7.0. Conversely, oligotrophic lakes were typically colonized by organisms with a pH optima below 7.0.[77] Thus, the eutrophic Yangtze River Flood Plain lakes of China are unlikely candidates for lake acidification in spite of the huge quantities of sulfur-rich coal which is consumed in this area of the world.

Xuanwu Lake is typical of many of the shallow hypereutrophic bodies of water in the east China flood plain. Lakes such as Xuanwu, which are near large cities, receive substantial quantities of nutrient-rich industrial and municipal discharges. Such lakes are undergoing progressive eutrophication and typically display very shallow photic zones (less than 1 m). These lakes are dominated by blue-green and green algae for the majority of the year.

5. Conclusions from Xuanwu Lake Study

In conclusion, it would appear that even though Xuanwu Lake was exposed to increasing concentrations of SO_2 and NO_x during the last 30 years, there was no indication that the lake was acidifying. This conclusion was based on the absence of an upcore increase in acidophilic diatom taxa and on the Nanjing Ministry of Environment records which indicated that the pH of Xuanwu Lake has not fallen during the last decade. Thus, Xuanwu Lake, like many shallow Yangtze River flood plain lakes in the Pacific basin of China, appears to have sufficient algal productivity and buffering capacity to prevent its acidification. Wind-blown, carbonate-rich material (e.g., loess) plus alkalinity generated by the aquatic vegetation in this extremely eutrophic (hypereutrophic) lake, render it an unlikely candidate for lake acidification.

IV. SUMMARY

The water in lakes (even those in granitic basins) does not behave like distilled water. When one stops pouring acid into a container of distilled water, its pH stabilizes and remains low and constant over time. By way of contrast, when one stops acidifying a lake, its pH does not remain low and constant but often rebounds in the direction of its preacidified condition if sulfate reduction occurs in its basin.

The sediment core chronology for Achterste Goorven and Xuanwu Lake support the conclusion that useful stratigraphic information on a decade-to-decade level of precision can be recovered from an analysis of even shallow lake sediments. Year-to-year analyses of shallow lake stratigraphic information is, however, impossible due to the degree of wind and biologically induced sediment mixing which takes place.

Lead-210 downcore chronologies deviate from exponentiality in lakes where sediment mixing occurs or in lakes where lead-210 outlet losses are high. Meromictic lakes or humic lakes with anaerobic hypolimnia possessing few invertebrates and minimal lead-210 loss from their entrained sediments are ideal systems for inferring downcore sedimentation rates.

Achterste Goorven was undergoing rapid acidification during the period 1950 to 1975. During the last decade, the rate of its diatom-inferred acidification was significantly reduced.

Lake 29B was undergoing rapid acidification until the late 1970s when the rate of its acidification was dramatically reduced. During the 1980s the lake began deacidifying.

Deacidification does not result from the simple cessation of acid addition. Instead, sulfate conversion to sulfide by *Desulphovibrio* sp. generates sufficient alkalinity in lakes impacted by acid rain to deacidify them as soon as the rate of sulfuric acid addition is substantially reduced.

A significant correlation between the density of sulfate-reducing bacteria and the magnitude of sulfur loading was observed.[58] This suggests that given sufficient organic matter, the greater the sulfate content in the soft water lakes, the greater the number of sulfate-reducing bacteria and the greater the alkalinity generated from sulfide production of *Desulphovibrio* spp.

Lakes which have acidified during the last 30 to 40 years have a characteristic assemblage of diatoms (microscopic algae). This assemblage is frequently dominated by acid-tolerant and trace metal-tolerant diatoms which are well preserved in the sediments of most acid lakes. Thus, it is possible to use the downcore changes in diatom-relative abundance to infer changes in paleo pH.

The erection of the 381-m high INCO ''superstack'' in 1972 has been shown to have significantly reduced the local deposition of sulfur, selected metals, and other pollutants released from the smelters in the Sudbury basin. The decrease in sulfur flux to the sediments of 150 lakes downwind of INCO, the largest point source of SO_2 emissions in the world, was studied to determine whether some of these lakes would maintain their depressed pH or whether their pH would ''bounce back'' following the reduction of SO_2 loading.

To determine whether man-induced lake acidification was rapidly reversible, 150 lakes north of Sudbury, Ontario, were cored during the summer of 1986. The cores from 2 of these 150 lakes were selected for fine sectioning because there was little if any evidence of bioturbation in the profundal sediments of these two lakes. The 2-mm thick sections from the shallower of these two lakes (Lake 29B) revealed a significant ($p < 0.05$) upcore increase in diatom-inferred pH from 4.3 at 6 mm (circa 1983) to 5.0 at 0 to 1 mm (circa 1986). This was attributed to the reduction in SO_2 loadings in the region around Sudbury during the last 6 years.

Lakes undergoing eutrophication are unlikely candidates for acidification. Presumably this is due to alkalinity generated during photosynthesis.

All else being equal, the larger the volume of a lake the more time it will require to acidify or deacidify.

Paleolimnological data based on a 16-cm long sediment core removed from near the center of Xuanwu Lake indicated that (nutrient) pollution-tolerant diatoms have replaced pollution-intolerant taxa which were common near the base of the core. This was consistent with the hypothesis that Xuanwu Lake is undergoing progressive eutrophication and not progressive acidification.

Xuanwu Lake, like many similar shallow polymictic lakes in eastern China, is not acidifying even though it is situated in the center of one of the most populous sulfur-rich coal burning areas of the world. Furthermore, its progressive hypereutrophication has not resulted in anoxia or any of the associated suite of problems experienced by deeper hypereutrophic lakes because Xuanwu Lake is polymictic.

V. APPENDIX

A. Problems with Diatom Stratigraphy in Shallow Lakes

The moorland pools of Europe and the Yangtze River flood plain lakes of China are very shallow (mean depth less than 2 m). Shallow ponds with little topographic relief are prone to sediment resuspension.[31] As a result, their entire sediment surface layer is subjected to wind-induced mixing. In addition, strong winds may deposit layers of loess (wind blown fines) diluting the authigenic materials deposited within these shallow lakes. Nevertheless, valuable stratigraphic information was collected from a moorland pool with a maximum depth of slightly less than 2 m (Achterste Goorven) and a shallow flood plain lake (Xuanwu Lake) with a maximum depth of 2.3 m.

Bioturbation from burrowing invertebrates may also distort sediment stratigraphy. The vertical and chronological resolution of stratigraphic studies can be severely limited by these disturbances.[78] Hakanson and Kallstrom[79] and Christensen[80] noted that each sample from a specific depth in a core contains materials of some *average* age, and that the distribution of ages around that average varies according to the degree of bioturbation and other mixing processes which have taken place at the core sites. Lead-210 dating reflects only this *average* age for each depth.[83]

Error in dating is a major problem in this field of research. Several authors, e.g., Robbins,[81] Oldfield and Appleby,[82] and Davis et al.[46] have discussed these problems. The counting error for radionuclides (^{137}Cs; ^{210}Pb) is only one of several sources of error in radionuclide dates. Of the many lake sediment cores dated by ^{210}Pb for lake acidification studies carried out by Geological Survey of Ontario, over half have ^{210}Pb profiles departing from exponentiality in the top few cm, suggesting[83] that a mixing zone has been formed due to some combination of bioturbation and physical process.

Direct evidence of bioturbation in lake sediments, in the form of the animals themselves or their burrows (i.e., tubificid worms, chironomid larvae, and clams) has often been observed.[30] Although it may be possible to minimize the effect of sediment mixing on chronological resolution by carefully selecting lakes whose sediments are undisturbed, this would eliminate most of the soft water, oligotrophic lakes impacted by acid rain. Thus, except for certain lakes (e.g., varved lakes, lakes with anaerobic bottom waters or ultra-oligotrophic lakes with negligible populations of bioturbators), we have no choice but to be satisfied with less than optimal chronological resolution, probably no better than ± 5 years, and quite possibly, typically ± 10 to 20 years.[83,84]

During periods of drought, shallow lakes and pools such as Achterste Goorven evaporate to near dryness. Such conditions of desiccation and oxidation result in poor preservation of many microfossils.[85] Such conditions also alter water chemistry and may directly affect the rate of lake acidification.[86] During such periods mammalian species including deer, moose, and man may disturb these sediments by walking through their soft sediments.[87]

Sediment coring of shallow ponds such as the moorland ponds is often impeded by a dense stand of submersed aquatic vegetation (e.g., mosses) growing on top of the sediments. If the vegetation is not removed before coring, the uppermost layers of sediment near the mud-water interface will be hopelessly distorted as the core passes down through the vegetation cover.[87]

Certain fish such as numerous species belonging to the carp family disturb the sediments of shallow lakes (e.g., Xuanwu Lake, China).

In summary, several major factors reduce the level of chronological resolution for shallow lakes: physical factors such as (1) wind-induced sediment resuspension, (2) aeolain deposits which dilute authigenic material, (3) bioturbation from mammals walking through the sediments, and (4) overlying aquatic plant cover which prevents sediment coring. Thus, in such shallow lakes, we must be satisfied with a temporal (chronological) resolution of plus or minus 10 to 20 years.[30]

B. Problems with Diatom Stratigraphy in Deep Lakes

In deep lakes with extremely flocculent sediments another group of problems confront stratigraphers. Unconsolidated, highly flocculent sediments with greater than 95% water content are encountered in Canadian Shield lakes of moderate to great depth. A gravity or piston corer entering the mud-water interface of these sediments sets up currents which may destroy the stratigraphic pattern of the flocculent uppermost layers of these sediments.[88,89]

The presence of gas bubbles generated in or near the redox interface may also alter the stratigraphic pattern of such sediments. It is for this reason that the sediments of the meromictic Pink Lake are homogeneous, as these sediments are filled with methane bubbles.[90] As these bubbles coalesce and move up through the sediments, they mix them causing a substantial deviation from exponentiality of the lead-210 profile for the sediments in Pink Lake.

Unless the core from a flocculent type of sediment is transported to the laboratory for sectioning with great care it will be partially mixed on arrival.[89] For example, if the temperature of the lake sediments differs from that in the laboratory or the vehicle in which the core is transported to the laboratory, convection currents will be generated which partially mix the sediments of the core unless the sediments are fairly well consolidated.

Core extrusion results in the "smearing" of that section of the core which comes in contact with the core liner. In very unconsolidated sediments this smearing action generates currents within the core which are capable of distorting sediments located even in or near the center of the core tube.[89]

In conclusion, few lakes are cored in which substantial downcore deviations from ^{210}Pb exponentiality do not occur. The reasons for this range from bioturbation, physical mixing, and gas bubble-induced mixing to sediment core distortion of flocculent sediments during their capture, transport, or extrusion.

C. Sulfur Cycle and Bacterial Sulfate Reduction in the Hydrosphere

Autotrophic and heterotrophic bacteria normally obtain their sulfur from sulfate in the environment. Since most of this sulfur is used to form amino acids containing S-H groups, it must undergo reduction during assimilative processes. The great majority of heterotrophic bacteria can liberate sulfur as H_2S from proteins and their decomposition products. Sulfate-reducing bacteria are anaerobic or facultative anaerobic and heterotrophic using sulfate as a hydrogen acceptor. It is nearly always possible to find some evidence of such reduction wherever sulfate and organic matter content coexist in the ecosystem in the absence of oxygen, and the hydrogen sulfide produced in the anaerobic hypolimnetic regions of lake or sediments is ultimately reoxidized to SO_4 completing the cycle. This can happen primarily involving bacteria; lake acidification processes result in an increased abundance of coarse organic matter on surface sediments and decreased aerobic microbial decomposition. However, greater abundance of sulfate reducers are triggered due to availability of excessive SO_4 and organic matter. This study has indeed indicated that bacterial sulfate consumption in the anaerobic and organic-rich lake sediments under acid stress has contributed to alkalinity generation as evidenced by increased abundance of alkalophilic diatoms.

ACKNOWLEDGMENTS

This research was a result of a team effort which attempted to combine expertise in paleolimnology, microbiology, and chemistry. We sincerely acknowledge each of the following for their valuable technical assistance: Mike Agbeti, John Ciolfi, Sushil Dixit, Pat Hayes and Kit Yung (Brock University, Biological Sciences Department), and A. A. Jurkovic (Ecotoxicology Section, Rivers Research Branch, National Water Research Institute, Burlington, Ontario). We would also like to acknowledge the advice and assistance of Dr. H.

Thode, McMaster University, and Drs. H. Wong, J. Nriagu, and M. Thompson of the Canada Centre for Inland Waters.

In addition, we wish to thank the Ministry of Natural Resources (Geological Survey Branch) for providing us with valuable background information and logistical support in the Sudbury area.

We are grateful to the Ontario-Jiangsu Educational Exchange program, Nanjing Normal University, and the Natural Sciences and Engineering Research Council of Canada (NSERC) for providing funds to carry out research on Xuanwu Lake in China. The Research Institute for Nature Management, P.O. Box 46, NL-3956-2R Leersum, The Netherlands and NSERC provided support for the Achterste Goorven Study. We are grateful to each of the above and to Bert Stahl, Doug Evans, and Carole Donaldson for respectively assisting in the sediment coring, lead-210 analyses, and pollen analyses of the Canadian Shield lakes. We are also grateful to Mary Roberts, Linda Vidican, Robb Irvine, and Elizabeth McCurdy for typing this manuscript.

REFERENCES

1. **Dixit, S. and Smol, J. P.**, Algal assemblages in acid-stressed lakes with particular emphasis on diatoms and chrysophytes, in *Acid Stress and Aquatic Microbial Interactions*, Rao, S. S., Ed., CRC Press, Boca Raton, FL, 1989, chap. 7.
2. **Gravat, L.**, Sulphate in precipitation as observed by the European atmospheric chemistry network, *Atmos. Environ.*, 12, 413, 1978.
3. **Wright, R. F. and Gjessing, E. T.**, Acid precipitation: changes in the chemical composition of lakes, *Ambio*, 5, 219, 1976.
4. Ontario Research Foundation, *The Removal of Sulphur Gases from Smelter Fumes*, King's Printer, Toronto, 1979.
5. **Gorham, E. and Gordon, A. G.**, The influence of smelter fumes upon the chemical composition of lake waters near Sudbury, Ontario and upon the surrounding vegetation, *Can. J. Bot.*, 38, 477, 1960.
6. **Fortescue, J., Dickman, M., Pitblado, R., Singhroy, V., Stahl, H., and Tanis, F.**, A study of Canadian acid lakes of the north (ASCALON), Geological Survey of Ontario Open File Report, Toronto, Ontario, 1988.
7. **Battarbee, R. W.**, Diatom analysis and the acidification of lakes, *Phil. Trans. R. Soc. London, Ser. B*, 305, 451, 1984.
8. **Hakala, I.**, 1971. A new model of the Kajak bottom sampler and other improvements in the zoobenthos sampling techniques, *Ann. Zool. Fen.*, 8(3), 422.
9. **Dickman, M., Hayes, P., and Fortescue, J.**, A comparison of two indices of diatom inferred pH, *J. Water Pollut. Res. Can.*, 20, 9, 1985.
10. **Smil, V.**, Energy development in China, *Energy Policy*, 9, 113, 1981.
11. **Zhao, D., Shiffen, M., Lenan, C., and Kena, L.**, Acid rain investigations in Beijing, *HYKX*, 2, 50, 1981.
12. **Rao, S. S., Paulini, D., and Leppard, G. G.**, Effects of low pH stress on the morphology and activity of bacteria from lakes receiving acid precipitation, *Hydrobiologia*, 114, 115, 1984.
13. **Zimmerman, R., Iturriaga, R., and Brick, J. B.**, Simultaneous determination of the total number of aquatic bacteria and the number thereof involved in respiration, *Environ. Microbiol.*, 36, 926, 1978.
14. **Smil, V.**, *The Bad Earth*, M. E. Sharpe, Armonk, NY, 1984, 245.
15. **Dutka, B. J.**, Methods for Microbiological Analysis of Water, Wastewaters and Sediments, Inland Waters Directorate, Scientific Operations Division, CCIW, Burlington, Ontario, 1978.
16. **Fortescue, J., Dickman, M., and Terasmae, J.**, Multidisciplinary follow-up of regional pH patterns in lakes north of Lake Superior, Ontario Geological Survey Open File Number 5483, 1984.
17. **Eakins, J. D. and Morrison, R. T.**, A new procedure for the determination of lead-210 in lake and marine sediments, *Int. J. Appl. Radiat. Isot.*, 29, 531, 1978.
18. **Dickman, M., Dixit, S., Fortescue, J., Barlow, R., and Terasmae, J.**, Diatoms as indicators of the rate of lake acidification, *Water Air Soil Pollut.*, 21, 375, 1984.
19. **Beaver, J.**, Apparent Ecological Characteristics of some Common Freshwater Diatoms, Ontario Ministry of Environment Report and Technical Support Section, Central Region, 1981.

20. **Camburn, K. E. and Kingston, J. C.,** The Genus *Melosira* from Softwater Lakes with Special Reference to Northern Michigan, Wisconsin and Minnesota, PIRLA Manuscript, University of Minnesota, Duluth, 1985.

21. **Charles, D.,** A new diatom species, *Fragilaria acidobiontica,* from acid lakes in northeastern North America, in *Diatoms and Lake Acidity,* Smol, J. P. et al., Eds., Junk, Boston, 1986, 356.

22. **Cleve-Euler, A.,** 1951-1955. Die Diatomeen von Schweden und Finnland, Kunglia Svenska Velenskapsakademien Avhandlingar, Almqvist adn Wiksells, Stockholm. 5, 2(1); 3(3); 4(1) 1, 4(5); 5(4).

23. **Foged, N.,** *Diatoms in New Zealand, the North Island,* J. Cramer, Struss and Cramer Publ., Hirschberg, Germany, 1979, 225.

24. **Hustedt, F.,** Bacillariophyta (Diatomea), in *Die Susswasser Flora Mitteleuropas,* Vol. 10, Pascher, Ed., Jena, East Germany, 1930, 446.

25. **Husedt, F.,** Systematische und Okalogische Unter suchungen uber die Diatomeen — Flora von Java, Bali, und Sumatra Mach der material der Deutschen Limnologischen Sunda — Expedition III, *Arch. Hydrobiol. Suppl.,* 16, 274, 1937-1939.

26. **Patrick, R. and Reimer, C. W.,** *The Diatoms of the United States,* Vol. 1, Monogr. Ser. 13, Academy of Natural Sciences of Philadelphia, Philadelphia, 1966, 688.

27. **Patrick, R. and Reimer, C. W.,** *The Diatoms of the United States,* Vol. 2, Monogr. Ser. 13, Academy of Natural Sciences of Philadelphia, Philadelphia, 1975, 213.

28. **Germain, H.,** *Flore des Diatomees,* Societe Nouvelle des Editions, Boubee, Paris, 1981, 444.

29. **Renberg, I. and Hellberg, T.,** The pH history of lakes in southwestern Sweden, as calculated from fossil diatom flora of the sediments, *Ambio,* 11, 30, 1982.

30. **Robbins, J. A.,** Lead-210 dating of lake sediments, abstract, *Bull. Ecol. Soc. Am.,* 66, 257, 1985.

31. **Moss, B.,** Algal and other fossil evidence for major changes in Strumpshaw Broad, Norfolk, England in the last two centuries, *Br. Phycol. J.,* 14, 263, 1979.

32. **Moss, B.,** Further studies on the palaeolimnology and changes in the phosphorous budget of Barton Broad, Norfolk, *Freshwater Biol.,* 10, 261, 1980.

33. **Coesel, P. F. M., Kwakkestein, R., and Verschoon, A.,** Oligotrophication and eutrophication tendencies in some Dutch moorland pools, as reflected in their Desmid flora, *Hydrobiologia,* 61, 21, 1978.

34. **Van Dam, H., Suurmond, G., and ter Braak, J. F.,** Impact of acidification on diatoms and chemistry of Dutch moorland pools, *Hydrobiologia,* 83, 425, 1981.

35. **Van Dam, H. and Beljaars, K.,** Nachweis von Versaurung in West-Eurpaischen Kalkarmen stehenden Gewassern durch Vergleich von alten und rezenten Kiesselalgenproben, in *Gewasserversauerung in der Bundesrepublik Deutsland,* Wieting, J., Lenhart, B., Steinberg, C., Hamm, A., and Schoen, R., Eds., Umweltbundesamt and Schmidt, Berlin, 1984, 184.

36. **Roelofs, J. G. M.,** Impact of acidification and eutrophication on macrophyte communities in soft waters in The Netherlands, *Aquat. Bot.,* 17, 139, 1983.

37. **Westoff, V., Bakker, P. A., Van Leeuwen, C. G., Van der Voo, E. E., and Zonneveld, I. S.,** Wilde planten, III, De hogre gronden, Vereniging tot Behoud van Natuurmonumenten in Nederland, Amsterdam, 1973.

38. **DeSmit, J. T.,** Heathland Vegetations of The Netherlands, Ph.D. dissertation, Rijksuniversiteit, Utrecht, Netherlands, 1975.

39. **Heimans, J.,** De Desmidiaceenflora van de Oisterwijksche Vennen, *Nederl. Kruik. Arch.,* 31, 245, 1925.

40. **Renberg, I.,** A sedimentary diatom record of severe acidification in Lake Blamissusjon, N. Sweden through natural soil processes, in *Diatoms and Lake Acidity,* Smol, J. P. et al., Eds., Dr. W. Junk, Dordrecht, Netherlands, 1986, 115.

41. Chemical Report from the Chemical Co-Ordinating Centre for the Second Phase of the Co-Operative Program for Monitoring and Evaluation of Long-Range Transmission of Air Pollutants in Europe (EMEP), EMEP/CCC Report, Norwegian Institute for Air Research (NILU), Lillestrom, Norway, 1984, 120.

42. KNMI/RIV, Chemical Composition of Precipitation over The Netherlands, Annual Report 1982, Joint KNMI/RIV project — Koninklijk Meteorologisch Instituut, De Bilt/Rijksinstituut voorde Volksgezondheid, Bilthoven, The Netherlands, 1983, 16.

43. **Berge, F.,** pH forandringer og sedimentasjon av diatomeer i Langtern. Norges Teknisk-Nataurvitenskapelige Forskningsgrad, Internal Report, Vol. 11, SNSF, Oslo, 1975, 1.

44. **Berge, F.,** Kieselarlgar og pH i noen innsjoer i Agder og Hordaland, S.N.S.F.-prosjeket IR, Vol. 42, SNSF, Oslo, 1979, 64.

45. **Tolonen, K. and Jaakkola, T.,** History of lake acidification and air pollution studied on sediments in south Finland, *Ann. Bot. Fennici,* 20, 57, 1983.

46. **Davis, R. B., Norton, S. A., Hess, T. C., and Brakke, D. F.,** Paleolimnological reconstruction of the effect of atmospheric deposition of acids and heavy metals on the chemistry and biology of lakes in New England and Norway, *Hydrobiologia,* 103, 113, 1983.

47. **Flower, R. J. and Battarbee, R. W.,** Diatom evidence for recent acidification of two Scottish lochs, *Nature,* 305, 130, 1983.

48. **Battarbee, R. W., Flower, R. J., Stevenson, A. C., and Pippey B.,** Lake acidification in Galloway: a paleolimnological test of competing hypothesis, *Nature,* 314, 350, 1985.

49. **Brugam, R. B. and Lusk, M.,** Diatom evidence for neutralization in acid surface mine lakes, in *Diatoms and Lake Acidity,* Smol, J. P. et al., Eds., Dr. W. Junk, Dordrecht, Netherlands, 1986, 115.

50. **LaZerte, B. D. and Dillon, P. J.,** Relative importance of anthropogenic natural sources of acidity of lakes and streams in central Ontario, *Can. J. Fish. Aquat. Sci.,* 41, 1664, 1984.

51. **Yan, N. D. and Miller, G. E.,** Effects of deposition of acids and metals on chemistry and biology of lakes near Sudbury, Ontario, *Adv. Environ. Sci. Technol.,* 15, 243, 1984.

52. **Hutchinson, T. C. and Havas, M.,** Recovery of previously acidified lakes near Coniston, Canada following reductions in atmospheric sulphur and metal emissions, *Water Air Soil Pollut.,* 28, 319, 1986.

53. **Forsberg, C., Morling, G., and Wetzel, R. G.,** Indications of the capacity for rapid reversibility of lake acidification, *Ambio,* 14, 164, 1985.

54. **Nriagu, J. O. and Rao, S. S.,** Response of lake sediments to change in trace metal emissions from smelters at Sudbury, Ontario, *Environ. Pollut.,* 44, 211, 1987.

55. **Schindler, D. W., Hesslein, H., Wageman, R., and Broecker, W. S.,** Effects of acidification on mobilization of heavy metals and radionuclides from the sediments of a fresh water lake, *Can. J. Fish. Aquat. Sci.,* 37, 373, 1980.

56. **Cook, R. B. and Schindler, D. W.,** The biogeochemistry of sulphur in an experimentally acidified lake, *Ecol. Bull.,* 35, 115, 1983.

57. **Harrison, A. G. and Thode, H. G.,** Mechanism of the bacterial reduction of sulphate from isotope fractionation studies, *Faraday Soc. Trans.* 54, 84, 1958.

58. **Thode, H. G., Dickman, M., and Rao, S. S.,** Effects on acid precipitation on sediment downcore profiles of diatoms, bacterial densities and sulphur isotope ratios in lakes north of Lake Superior, *Arch. Hydrobiol.,* 74, 397, 1987.

59. **Schnoor, J. L. and Stumm, W.,** Acidification of aquatic and terrestrial systems, in *Chemical Processes in Lakes,* Stumm, W., Ed., John Wiley & Sons, London, 1985, 311.

60. **Schiff, R.,** personal communication.

61. **Stahl, H.,** Influence of Sediment Redox Status on Lake Acidification in Northeastern Ontario, M.Sc. thesis, Laurentian University, Sudbury, Ontario, 1986.

62. **Hutchinson, G.,** *A Treatise on Limnology,* John Wiley & Sons, New York, 1957, 1015.

63. **Dillon, P. J., Scholer, P. J., and Evans, H. E.,** Lead-210 fluxes in acidified lakes, in *Sediments and Water Interactions,* Sly, P., Ed., Springer Verlag, Berlin, 1986, 491.

64. **Hesslein, R., Broecker, W. S., and Schindler, D. W.,** Fates of metal radiotracers added to a whole lake: sediment-water interactions, *Can. J. Fish Aquat. Sci.,* 37, 378, 1980.

65. Environment Canada, Acid Rain Reductions in Canada, Environment Canada Annual Report for 1985-1986, Ottawa, Canada, 1986.

66. **Shi, C. H.,** A comprehensive investigative study of lakes in the southern part of Jiangsu Province, *Proc. Natl.* Geographic Symposium of 1960, Science Press, Beijing, China, 1962, 227.

67. **Shi, C. H. and Liang, R. J.,** Lake Tai: the limnology of a shallow lake in China, Geo. J., 14, 319, 1987.

68. Nanjing Institute of Geography, Lacustrine Features and Progress of the Research Work in China, Limnological Group of the Nanjing Institute of Geography, Academia Sinica Report Series, Nanjing, China, 1981.

69. **Wong, H. T., Tao, H. S., Woing, S. J., and Zhang, L.,** *Chinese Lakes,* Commerce Publisher, Bejing, China, 1984, 247.

70. **Chang, W. B.,** Large lakes of China, *J. Great Lakes Res. Int. Assoc. Great Lakes Res.,* 13, 236, 1987.

71. **Herdendorf, C. E.,** Large lakes of the world, *J. Great Lakes Res.,* 8, 370, 1982.

72. **Ho, G. Z.,** Jiangsu main lake water bodies, *Nanjing Normal Univ.,* 1, 25, 1983.

73. **Xu, J. Z., Dickman, M. D., Wang, L. P., Shi, G. X., Xu, X. S., and Zhou, K. Y.,** The Limnology and Paleolimnology of a Shallow Eutrophic Lake in Eastern China, Nanjing Normal University, Nanjing, China, 1988, 37.

74. **Till, B.,** *In Search of Old Nanjing,* Joint, Hong Kong, 1984, 242.

75. **Seenayya, G.,** Ecological studies in the plankton of certain freshwater ponds of Hydraalad, India, *Hydrobiologia,* 39, 247, 1972.

76. **Liu, Y. I.,** A Survey of the Planktonic Algae and their Relation to the Pollution of Water Bodies in Xuanwu Lake, Nanjing Teacher's College, Nanjing, China, 1981.

77. **Berzins, B. and Pejler, B.,** Rotifer occurrence in relation to pH, *Hydrobiologia,* 147, 107, 1987.

78. **Robbins, J. A., McCall, P. L., Fischer, J. B., and Krezoski, J. R.,** Effect of deposit feeders on migration of (137) Cs in lake sediments, *Earth Planet. Sci. Lett.,* 42, 277, 1979.

79. **Hakanson, L. and Kallastrom, A.,** An equation of state for biologically active lake sediments and its implications for interpretations of sediment data, *Sedimentology,* 25, 205, 1978.

80. **Christensen, E. R.,** A model for radionuclides in sediments influenced by mixing and compaction, *J. Geophys. Res.,* 87, 566, 1982.

81. **Robbins, J. A.,** Geochemical and geophysical applications of radioactive lead, in *The Biogeochemistry of Lead in the Environment,* Nriagu, J. O., Ed., Elsevier, Amsterdam, 1978, 285.

82. **Oldfield, F. and Appleby, P. G.,** Empirical testing of ^{210}Pb dating models for lake sediments, in *Lake Sediments and Environmental History,* Haworth, E. Y. and Lund, J. W. G., Eds., University of Minnesota Press, Minneapolis, 1984, 93.

83. **Davis, R. B. and Smol, J. P.,** The use of sedimentary remains of siliceous algae for inferring past chemistry of lake-water problems, potential and research needs, in *Diatoms and Lake Acidity,* Smol, J. P., et al., Eds., Dr. W. Junk, Dordrecht, Netherlands, 1986, 191.

84. **Binford, M. W.,** Critical assessments of methods of dating stratigraphic events in the past few hundred years, *Abstr. Bull. Ecol. Soc. Am.,* 66, 142, 1985.

85. **Iversen, J.,** Retrogressive development of a forest ecosystem demonstrated by pollen diagrams from fossil mor, *Oikos,* 12 (Suppl.), 35, 1969.

86. **Vangenechten, J. H. D., Bosmans, F., and Deckers, F. B. H.,** Effects of short-term changes in rainwater supply on the ionic composition of acid moorland pools in the Campine of Antwerp (Belgium), *Hydrobiologia,* 76, 149, 1981.

87. **Dickman, M., van Dam, H., van Geel, B., Klink, A., and van der Wijk, A.,** Acidification of a Dutch moorland pool: a paleolimnological study, *Arch. Hydrobiol.,* 109, 377, 1987.

88. **Diamond, M. L., Mackay, D., Phalp, D., Ahier, B., Landsberger, S., and Cornett, R. J.,** Application of the fugacity/activity model to predicting the behaviour of arsenic in lakes, Technology Transfer Conference Proceedings — Water Quality Research, Toronto, Ontario, 1987.

89. **Chant, L.,** Sediment isotope stratigraphy in a core from Perch Lake containing an unconsolidated flocculent layer (0-4 cm), SCI Annual Meeting, (Society of Canadian Limnologists, January 6, 1988.

90. **Dickman, M.,** unpublished data.

Chapter 9

PROTOZOAN BACTERIVORY IN ACIDIFIED WATERS: METHODS OF ANALYSIS AND THE EFFECT OF pH

Sarah C. Tremaine and Aaron L. Mills

TABLE OF CONTENTS

I. INTRODUCTION

The acidification of aquatic ecosystems due to acid rain or acid mine drainage has caused extensive reductions in fish and mollusk populations. Lakes and streams devoid of fish suffer trophic reorganization where predatory invertebrates assume the top predator niche with concomitant shifts in zooplankton, protozoan, bacterial, and algal communities. This chapter will focus on the effect of acid stress on protozoans.

The role of protozoans as bacterivores in aquatic ecosystems has been described by many authors.[1-9] Bacteria decompose 10 to 50% of the total fixed carbon in marine water column systems,[10-12] and 40 to 80% in lake ecosystems.[13] Estimates of bacterial carbon conversion efficiency are 40 to 80%, so a considerable quantity of carbon and nutrients (N, P, S, etc.) are available to higher trophic levels which consume bacteria. Consumption of bacteria has been documented in protozoans,[5] amphipods,[14] rotifers,[15] prosobranchs,[16,17] and bivalves.[16,18,19] It has been hypothesized that bacterivorous protozoans (ciliates, amoebae, and flagellates)[20] are the dominant consumers of bacteria,[21] and therefore represent a critical link in energy transfer through aquatic food webs.

The effect of acidification on bacterivorous protozoans has been virtually ignored in acidification studies, with the notable exceptions of research on several Florida, Ontario, and Michigan wetlands and lakes.[20,22-26] These studies examined the possible correlation between protozoan abundance, biomass, and colonization rates and pH (reviewed in detail below). Presented here are the first data on protozoan grazing rates under acid stress. The determination of protozoan grazing rates is an area of research still under development. These methods and their associated limitations have not been reviewed previously.

II. METHODS OF STUDY

Two approaches have been used to quantify bacterivory in aquatic systems: (1) labeling of food particles and (2) calculation of grazing from the growth of prey communities when grazing pressure is relieved by selective chemical inhibition, filtration, or dilution. These methods are currently being evaluated and we review the progress of this research below.

A. Labeled Particles

Some investigators have used "inert" particles as surrogate prey for bacterivory. Such particles are of three kinds: (1) inert fluorescent particles the size of bacteria, (2) radiolabeled bacteria, and (3) fluorescent-labeled bacteria (FLB). The critical assumptions of these methods are that the label is a permanent marker of the particle and that the particles are consumed at the same rate as the bacteria they mimic, i.e., without bias.

1. Fluorescent Microspheres
a. Introduction
Fluorescent microspheres are inert polymeric spheres impregnated with fluorescent dyes and are commercially available in many sizes. The most commonly used size is 0.6 μm in diameter; these are closest in size to natural bacteria in many aquatic systems. The fluorescent dyes are permanent markers which show little fading under microscopic examination. Microspheres are generally added at concentrations between 10 and 40% of the bacterial concentration. Most researchers use incubation times of 10 to 20 min.

Methods of preservation vary, but recent reports by Sieracki et al.[27] and Sanders[28] indicate that some preservatives cause the ejection of ingested material (see below). Early reports suggest that grazing bias varies from system to system. Jonsson[29] reported no grazing bias, while other researchers reported bias factors of 1.32 ± 0.11,[30] 1.5,[31] and 4 to 10[32] against the microspheres. Our research, reported below, indicates that there may be a threshold

bead concentration below which grazing bias increases dramatically. Finally, some researchers have had difficulty observing any bead ingestion in some systems.[33]

b. Microsphere Experiments

Fluorescent microspheres (or beads) have been used to quantify protozoan grazing rates
in our laboratory experiments on protozoan cultures derived from a small eutrophic pond
near the University of Virginia. Two experiments were conducted to determine bacterial
grazing mortality (reduction in numbers due to grazing) and grazing bias toward the beads,
i.e., whether the beads were incorporated at the same rate as bacteria. Protozoan cultures
were established in 10x Osterhour's solution with sterile rice grains.[34]

The first experiment was conducted to determine whether the grazers showed a microsphere
grazing bias. Two subcultures were established from a single original culture to ensure that
the same protozoan community was used in each experiment. These subcultures had high
and low concentrations of protozoans at the initiation of the grazing experiment: 166 ± 19
(1 SD) grazers per milliliter and 2951 ± 21 grazers per milliliter. Bacterial concentrations
were 2.50×10^6 and 3.14×10^6 per milliliter ($\pm 10\%$ counting error) in the low and high
treatments, respectively. The experiments were conducted in 600-ml sterile beakers with
300 ml of culture in each of three beakers. Fluoresbrite fluorescent microspheres (Polysciences, Inc.) 0.61 ± 0.01 µm in diameter were added to the beakers in the following
concentrations: 10.3, 19.1, and 29.7% of the bacterial concentration for the low grazer
concentration treatment and 8.9, 15.6, and 28.9% of the bacterial concentration for the high
protozoan concentration treatments. Subsamples were taken over time, preserved with 10%
buffered glutaraldehyde (1% final concentration; v/v), and protozoans and ingested beads
were enumerated within 2 d using the primulin staining method on 1.0-µm pore size Nuclepore filters.[35] Bacteria were enumerated by the acridine orange direct count method.[36]
Beads were enumerated on irgalin black-stained 0.2-µm diameter pore size Nuclepore filters.

To determine particle retention times, and hence the period over which the grazing rate
should be evaluated, the experiments were run for 30 min and the data were analyzed by
the method of McManus and Fuhrman.[31] They hypothesized that the concentration of beads
in a grazer at any time, C_t, is dependent on the maximum bead concentration per grazer
(C_m) minus a bead concentration factor dependent on the egestion rate and time:

$$C_t = C_m[1 - e^{-kt}] \qquad (1)$$

where k is the particle egestion rate (per units of time). Plots were made of ln C_t vs. time
to determine ln C_m by inspection. Then, a least squares regression on ln $(C_m - C_t)$ vs. time
was run. The inverse of the slope of the regression line is l/k, the particle retention time.
Using this analysis, the particle retention time was determined to be 14 min $\pm 5\%$, and 10
min $\pm 3\%$ in the low and high grazer concentration treatments, respectively. The particle
retention time set the limits of the time interval over which the particle ingestion rates were
calculated.

Total particle ingestion rates were calculated from the least squares regression of number
of beads ingested per grazer vs. time (Figures 1 and 2) and corrected by the concentration
of beads relative to the bacteria to give the bacterial grazing rate. The bead grazing rate was
then added to the bacterial grazing rate to give the total particle ingestion rate (Table 1).
To evaluate a bead grazing bias factor, the following assumption was made: if there is no
bias, the ratio of total particle ingestion rate at one bead concentration to another should be
equal to the ratio of the bead concentrations (as a percent of the bacterial concentration;
Table 1). Note that deviation from equality in this relationship would have to exceed total
counting error (10 to 20%) to be considered a bias.

Total particle ingestion rates for the experiment with a low concentration of protozoans

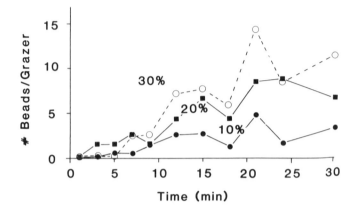

FIGURE 1.   Protozoan grazing (low numbers of protozoa) on fluorescent microspheres provided at an approximate percentage of the bacterial numbers present in the suspension. See text for actual percentages.

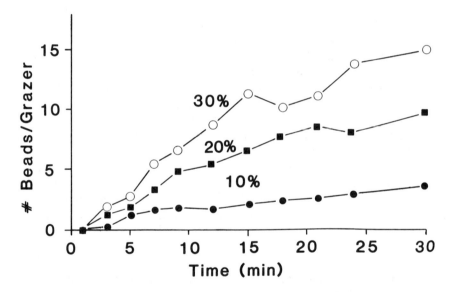

FIGURE 2.   Protozoan grazing (high numbers of protozoa) on fluorescent microspheres provided at an approximate percentage of the bacterial numbers present in the suspension. See text for actual percentages.

were 2.60, 2.59, and 2.65 particles per grazer per minute at 10.3, 19.1, and 29.7% beads, respectively. Bead ingestion biases were $+8$ and $+7\%$ ("$+$" being bias for the beads, "$-$" being bias against the beads; Table 1). Thus, a bias is present, but insignificant in terms of the overall error of the analysis. Total particle ingestion rates for the experiment with a high concentration of protozoans were 1.54, 3.54, and 3.43 particles per grazer per minute at 8.9, 15.6, and 28.9% beads, respectively. Bead ingestion biases were -53%, $+15\%$. Although the -53% value is significant relative to the error of the analysis, the bias was due to the extremely low bead ingestion rate for the lowest bead concentration treatment. This treatment provided an unrealistically low number of labeled particles with respect to the nutritional requirements of the grazers (but see further discussion below).

Clearance rates (the volume of suspension cleared by a single grazer in 1 h) were also calculated from the total particle ingestion rates and bacterial concentration. The rates were 0.03 to 0.05 µl per grazer per hour over all of the experiments (Table 1). The constancy

Table 1
SUMMARY OF THE FLUORESCENT MICROSPHERE BIAS EXPERIMENT

Protozoan conc (ml^{-1})	% Beads (conc, ml^{-1})	Ingestion rate (grazers per minute)		Bias (%)	Clearance volume (microliters per grazer per hour)	Available number (per grazer)	
		Bead	Particle			Beads	Particles
Low	10.3	0.243	2.60		0.04	2181	2.35 × 10^4
166	(3.62 × 10^5)			+8			
	19.1	0.416	2.59		0.05	3723	1.93 × 10^4
	(6.18 × 10^5)			+7			
	29.7	0.607	2.65		0.03	5753	3.27 × 10^4
	(9.55 × 10^5)						
High	8.9	0.126	1.54		0.03	98	0.12 × 10^4
2951	(2.88 × 10^5)			−53			
	15.6	0.477	3.54		0.04	233	0.19 × 10^4
	(6.87 × 10^5)			+15			
	28.9	0.770	3.43		0.03	345	0.20 × 10^4
	(10.19 × 10^5)						

Note: A positive bias (+) is a bias in favor of the microspheres and a negative bias (−) is against the microspheres.

of clearance rate, despite the −53% grazing bias in the high protozoan, low bead treatment, led us to compare the concentrations of beads and total particles per grazer (Table 1). The number of beads per grazer in the low protozoan concentration experiment was an order of magnitude greater than in the high protozoan concentration experiment (Table 1). The lowest concentration was 97 beads per grazer in the high protozoan/low bead treatment; which, coincidentally, had the −53% bias. The total number of particles per grazer was more constant over the treatments (Table 1), but the high protozoan/low bead treatment had the lowest total particle per grazer concentration (1.2 × 10^3). This suggests that there is a threshold concentration (200 beads per grazer or 700 beads per microliter) below which there is a grazing bias against the beads. No significant grazing bias for or against the beads was evident at bead concentrations above that threshold.

In the second experiment, we compared the grazing rates of protozoans of different sizes in suspension together: a *Paramecium* sp. (approximately 100 µm long) and an unidentified oligotrichous ciliate (approximately 10 µm long). The concentration of the *Paramecium* sp. was 355 cells per milliliter and that of the ciliate was 1.09 × 10^3 cells per milliliter, making the total grazer concentration 1.45 × 10^3 cells per milliliter. The bacterial concentration in the culture was 6.86 × 10^6 cells per milliliter. A 15-min grazing experiment was conducted in 600-ml beakers with 200 ml of suspension in each of six beakers. Fluoresbrite (Polysciences) fluorescent microspheres (0.61 ± 0.01 µm diameter) were added to the beakers at either 2.1, 5.3, or 10% of the bacterial concentration. Subsamples were taken over time and preserved with glutaraldehyde to yield a final concentration of 1% (v/v).[35] Bacteria and beads were enumerated on 0.2-µm pore size Nuclepore filters.[36] Protozoans and beads within the protozoans were enumerated after staining with primulin and filtration of the particles onto 1.0-µm pore size Nuclepore filters.[35] For each of 24 subsamples, 24 fields were counted at 250× magnification. Ingestion rates were below the limit of detection (due to the counting procedure) in the 2 and 5% experiments; data presented here are from the 10% bead experiment. The number of beads ingested per grazer has been plotted for both the *Paramecium* sp. and the ciliate (Figure 3), and a least squares regression provided the bead grazing rates. The *Paramecium* grazing rate was 2.43 beads per grazer per minute, which is equivalent to 5.69 × 10^5 bacteria per milliliter per hour, assuming no

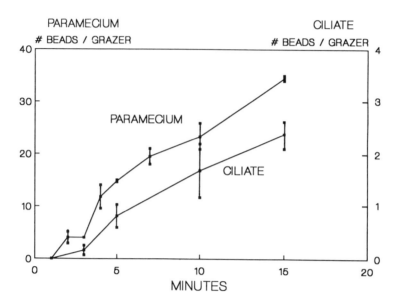

FIGURE 3.   Fluorescent microsphere consumption with time by *Paramecium* sp. and an unidentified ciliate in coculture. Error bars represent 1 SE.

Table 2
PROTOZOAN GRAZING AND CLEARANCE RATES FOR LABELED PARTICLES

Type of particle	Organism	Grazing rate (bacteria per grazer per hour)	Clearance rate (nanoliters per grazer per hour)	Ref.
Microsphere	*Dinobryon*	0.6	140	30
	Ciliates		1000	30
	Rotifers		14000	30
	Flagellates	0.04—0.4	0.6—1.9	31
	Epistylis rotanus	0.2—2.0	230—1260	110
	Strombidium sp.	0.2—2.0	260—900	110
	Flagellate		2—40	111
	Ciliate	0.6—0.65		112
	Flagellates and ciliates	1.5—3.5	30—50	This study
	Paramecium sp.	24.5	200	This study
	Ciliate sp.	1.8	15	This study
Fluorescent bacteria	Scuticociliates		140	32
	Oligotrichs		260	32
	Flagellates		3.2	32
Radiolabeled bacteria	Dinoflagellate spp.		400—8000	38
	Ciliates		700—11100	38
	Flagellates		10—65	38

grazing bias (Table 2). The ciliate grazing rate was 0.177 beads per grazer per minute, which is equivalent to 1.27×10^5 bacteria per milliliter per hour assuming no grazing bias.

In summary, protozoan grazing rates on bacteria have been estimated using fluorescent microspheres. These experiments indicate that there is no significant bias for or against the beads at concentrations greater than 200 beads per grazer or 700 beads per microliter. Grazing rates varied depending on the size of the predator: 5.69×10^5 bacteria per milliliter per hour for a 100-μm long *Paramecium* and 1.27×10^5 bacteria per milliliter per hour for a 10-μm long ciliate. This method of determining protozoan grazing rates using inert particles appears to be quite promising, but grazing bias must be evaluated for each system.

2. Radiolabeled Bacteria

Rates of protozoan grazing on bacteria have been quantified by measuring the accumulation of radioactivity in the grazers as they feed on radiolabeled bacteria. The compounds most often used to label bacteria have been ^{32}P-phosphate,[37] ^{3}H-amino acids,[38-40] and ^{14}C-glucose.[39] Clearance rates for bacterial predators range from 10 to 11,000 nl per individual per hour using bacteria labeled with ^{3}H-thymidine[38] (Table 2). Additionally, Ducklow et al.[41] reported that protozoan bacterivores consumed 50 to 100% of daily bacterial production in Loch Ewe, Scotland. The use of radiolabeled bacteria shows promise, but definitive studies of loss of activity by respiration and ingestion have not been published. Given appropriate testing, the use of radiolabeling may become a method of choice for short incubations.

3. Fluorescent Bacteria

Recently, Sherr et al.[32] reported the results of experiments using FLB and fluorescent microspheres to quantify protozoan grazing on bacteria. They found that pelagic oligotrichous ciliates and phagotrophic flagellates ingested FLB (Table 2) at 4 to 10 times the frequency of the rate for fluorescent microspheres. Sanders et al.[42] found no difference between ingestion rates of FLB and fluorescent microspheres. Sieracki et al.[27] observed that preservatives can cause the ejection of ingested material, confounding results from FLB and microsphere experiments; in contrast, Sanders[28] reports that cold glutaraldehyde (2% final concentration) did not produce a significant difference between ingestion of beads vs. fluorescent bacteria. Continued research will provide answers to the issues that this new technique raises.

B. Removal of Grazing Pressure

1. Selective Inhibition

a. Eukaryote Inhibition

Measurement of the grazing rate with a selective eukaryote inhibitor assumes that upon inhibition of grazers, observed bacterial growth is equal to the grazing rate.[43,44] This is predicated on two other assumptions: (1) that the target heterotrophic eukaryotes are immediately inhibited and (2) that nontarget photosynthetic eukaryotes and prokaryotes are unaffected by the eukaryote inhibitor. The most commonly used eukaryote inhibitors are cycloheximide,[1,32,39,43-50] thiram,[1,32,46,47] and colchicine.[46,47] Other inhibitors which have had limited use are demicolchicine,[46] neutral red,[47] griseofulvin,[47] and amphotericin B.[45] Each of these inhibitors violates the assumptions of selective inhibition by producing inconsistent inhibition of eukaryote grazers (Table 3),[45,52] and/or inhibiting nontarget microorganisms (Table 4).[52] These results preclude the use of these eukaryote inhibitors in quantitative grazing experiments.[45,47,52]

b. Prokaryote Inhibition

Prokaryote inhibitors have also been used in quantitative grazing experiments.[1,2,39,43,45,46,48-50,53-56] Measurement of grazing rates with a prokaryote inhibitor assumes that a decrease in the number of nonreproducing bacteria is quantitatively equal to the grazing rate of the protozoans in the sample water.[43,53,55] In this case, the eukaryotes are assumed to be uninhibited and to continue to graze while bacterial growth is inhibited. This method has three assumptions. First, the antibiotic inhibits the bacteria without altering the bacteria or their desirability as a food item. Antibiotics inhibit bacteria by interfering with cell wall synthesis, membrane function, protein synthesis, nucleic acid metabolism, or intermediary metabolism.[57] Antibiotics affecting cell wall synthesis, ribosomal function, and membrane function have been used in grazing experiments (Table 5). The drugs which affect cell wall synthesis cause cell lysis, which could be confounded with bacterial mortality due to grazing. Antibiotics affecting the cell membrane can result in cell leakage of metabolites which could enhance bacterial growth.

Table 3
EFFECT OF EUKARYOTE INHIBITORS ON TARGET EUKARYOTES

Drug	Organism(s)	Conc (mg·l^{-1})	Inhibition	System	Ref.
Cycloheximide	Field sample	100—200	Y	Soil	1
	Field sample	500	Y	Marine	44
	Field sample	100	Y	Marine	54
	Cyclidium sp.	2—56	N	Marine	47
	Cyclidium sp.	281	Y	Marine	47
	Culture	25—200	N	Freshwater	52
	Paramecium bursaria	1000	N	Laboratory	45
	P. caudatum	1000	N	Laboratory	45
	Cyclidium sp.	100	N	Freshwater	45
	Colpoda sp.	100	Y	Freshwater	45
	Isolate: flagellate	100—200	Y	Marine	46
	Isolate: ciliate	100—200	N	Marine	46
Thiram	*Cyclidium*	56—281	Y	Marine	47
	Field sample	10—200	Y	Soil	1
	Field sample	1	Y	Marine	54
	Isolate: flagellate	1—100	Y	Marine	46
	Isolate: ciliate	1—100	Y	Marine	46
	Field sample	1—100	Y	Marine	46
	Isolate: flag and ciliate	1	Y	Marine	37
Colchicine	Isolate: flagellate	25—200	N	Marine	46
	Isolate: ciliate	25—200	N	Marine	46
	Cyclidium sp.	56	N	Marine	47
	Cyclidium sp.	281	Y	Marine	47

Note: Inhibition greater than 90% = Y (yes) and less than 90% = N (no). System is the ecosystem type in which the research was conducted.

Table 4
EFFECT OF EUKARYOTE INHIBITORS ON NONTARGET MICROORGANISMS

Drug	Organism	Conc (mg·l^{-1})	Inhibition	System	Ref.
Cycloheximide	Phytoplankton	Not given	Y	Marine	48
	Chlorophyte	50	Y	Marine	47
	Crysophyte	50	Y	Marine	47
	Diatom (three species)	50	Y	Marine	47
	Phytoplankton	28	Y	Marine	39
	Synechococcus sp.	50—100	N	Marine	50
	Thiobacillus ferrooxidans	0.1—200	N	Freshwater	52
	Escherichia coli	Not given	N	Laboratory	51
	Bacterioplankton	1—100	N	Marine	54
	Bacterioplankton	50	Y	Marine	47
	Bacterioplankton	100	Y	Marine	46
	Anaerobic bacteria	50—100	Y	Freshwater	52
	Aerobic bacteria	0.1—200	N	Freshwater	52
Thiram	Photosynthetic eukaryotes	Target inhibited	Y	Marine	47
	Bacteria	10—100	Y	Marine	54
	Bacteria	Target inhibited	Y	Marine	47
	Bacterioplankton	1—100	Y	Marine	46
	Bacterioplankton	1	N	Marine	37
Colchicine	Photosynthetic eukaryotes	Target inhibited	Y	Marine	47
	Bacterioplankton	25—200	N	Marine	46

Note: Inhibition greater than 90% = Y (yes) and less than 90% = N (no). System is the ecosystem type in which the research was conducted.

Table 5
SITE OF ACTION FOR PROKARYOTE INHIBITORS USED IN GRAZING STUDIES

Inhibitor	Action	Ref.
Inhibits Cell Wall Synthesis		
Ampicillin	Cidal	50
Cephapirin		46, 54
Penicillin G	Cidal, gm +, aerobes lysis, acid sensitive	43, 45, 46, 48, 55
Vancomycin	16S RNA, methanogens not inhibited	46
Inhibits Ribosomal Function		
Chloramphenicol	Static, 70S	39, 45, 46, 54, 55
Gentamycin	30S	53
Neomycin	30S	55
Streptomycin	Cidal, 30S	1, 2, 48, 49, 55, 56
Inhibits Membrane Function		
Monensin	Na + ionophore	48
Valinomycin	K + ionophore	48

Table 6
THE EFFECT OF PROKARYOTE INHIBITORS ON TARGET PROKARYOTES

Drug	Organism	Conc $(mg \cdot l^{-1})$	Inhibition	System	Ref.
Chloramphenicol	Bacterioplankton	1—100	Y	Marine	46, 54
	Bacterioplankton	Not given	I	Freshwater	45
	Bacterioplankton	50	I	Marine	55
Cephapirin	Bacterioplankton	1—100	N	Marine	46, 54
Streptomycin	Plankton <203 μm	1—100	I	Marine	48
	Plankton <203 μm	1000	Y	Marine	48
	Bacterioplankton	50	I	Marine	55, 56
Monensin	Plankton <203 μm	1—100	I	Marine	48
	Plankton <203 μm	1000	Y	Marine	48
Valinomycin	Plankton <203 μm	1—10	I	Marine	48
Penicillin	Anaerobic bacteria	100	I	Freshwater	45
	Bacterioplankton	1—50	I	Marine	46, 55
	Bacterioplankton	100	Y	Marine	46
Vancomycin	Bacterioplankton	100—200	I	Marine	46
Neomycin	Bacterioplankton	50	I	Marine	55

Note: Inhibition greater than 90% = Y (yes), inhibition between 90 and 10% = I (incomplete), and inhibition less than 10% = N (no). System is the ecosystem in which the research took place.

The second assumption is that all of the bacteria are inhibited equally; however, many antibiotics inhibit some bacteria and not others (Table 6). Penicillin inhibits Gram-positive bacteria and is less effective on anaerobes[45] and Gram-negative bacteria (the latter dominate aquatic environments). Vancomycin does not inhibit methanogens because it acts to inhibit 16S RNA which is not present in methanogens.[58] Inhibition of nontarget microorganisms is also a problem with the use of prokaryote inhibitors (Table 7). Researchers have reported

Table 7
THE EFFECT OF PROKARYOTE INHIBITORS ON NONTARGET EUKARYOTES

Drug	Organism	Conc (mg l^{-1})	Inhibition	System	Ref.
Chloramphenicol	Heterotrophic	1	N	Marine	54
	eukaryotes	10	I	Marine	54
		50	Y	Marine	54
	Protozoan isolates	100	I	Freshwater	45
Penicillin	Protozoan isolates	100	I	Freshwater	45
Gentamycin	Photosynthetic microorganisms	40—250	I	Marine	53

Note: Inhibition greater than 90% = Y (yes), inhibition between 90 and 10% = I (incomplete), and inhibition less than 10% = N (no). System is the ecosystem in which the research took place.

incomplete inhibition of bacteria by chloramphenicol and penicillin (the most commonly used antibiotics) at concentrations where nontarget organisms are inhibited.[47]

Finally, the third assumption is that inhibition of a microorganism is constant and that the effect is immediate. Yetka and Wiebe[55] have shown that bacterial susceptibility to antibiotics is dependent on the growth phase of the bacteria. Bacteria in late log phase and stationary phase exhibit the greatest resistance to antibiotics, while bacteria in early log-phase growth exhibit the greatest antibiotic susceptibility. This would also introduce a time lag before inhibition occurs. These problems preclude the use of prokaryote inhibitors in quantitative grazing studies.

2. Filtration
a. Introduction
The filtration method of estimating grazing mortality eliminates eukaryote predators from water samples by filtering out organisms larger than the nominal pore size (usually 1 or 3 μm). Problems with the filtration method arise if there is a large proportion of the bacterial community attached to particles and therefore trapped on the filter.[7] This results in both elimination of grazing mortality and biased estimates of bacterial growth rates representative of only the nonparticle-associated bacteria. A second problem with the filtration method is that grazers in marine systems have been reported to pass through 0.4- and 0.6-μm pore size filters.[43,59] These microflagellates continue to graze and thus bias the estimate of grazing mortality.[59] However, if small (<1.0 μm) grazers constitute only a very small percentage of the total protozoan community, the bias may be insignificant in the 1st 24 h of an experiment. This method appears to be reliable when attached bacteria and small grazers are not numerically dominant.[7,60]

b. Filtration Experiments
To determine protozoan grazing rates using the filtration technique, two 10-l water samples were collected from the sediment-water interface of Lake Anna, VA, from two stations of with different pHs (4.39 and 5.24), and returned to the laboratory in insulated containers. Samples of the lake water were filtered through 1-μm pore size filters (Nuclepore) or not filtered and incubated for 24 h at *in situ* temperatures in 2-l flasks. Subsamples were taken from the flasks and preserved at 6-h intervals. Bacteria were enumerated using the acridine orange direct count method.[36] Protozoan grazing rates are calculated from comparison of bacterial growth rates in the presence (control, no filtration) and absence of grazing (grazers filtered out). Protozoan grazers were effectively removed by filtration (i.e., below the counting limit of detection). Changes in bacterial number over time in the 1-μm treatment

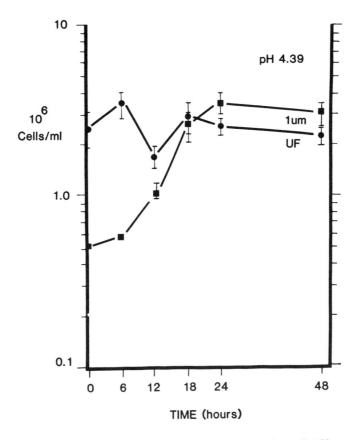

FIGURE 4. Estimation of protozoan grazing on bacteria at pH 4.39 as determined by the filtration method. Actual grazing rates were calculated on the basis of the regression of lnY vs. X for the 1st 24 h of the incubation. UF (●) represents unfiltered controls; 1 μm (■) represents suspensions that had been passed through a 1-μm pore size filter. Error bars represent 1 SE of the mean.

(Figures 4 and 5) were used to calculate bacterial grazing mortality by protozoans: 0.1061 and 0.0651 h^{-1} for the low and high pH stations, respectively. These rates are comparable to rates determined by other methods in the same system (see discussion below, Figure 6, Table 2, and Table 8). Filtration removed up to 50% of the bacterial community, (see time zero bacterial concentration, Figure 6) implying that a large proportion of the bacterial community was particle attached. It has been established that a large portion of the suspended bacteria in the bottom waters of Lake Anna are associated with particles; thus, filtration loss of bacteria was not surprising. This observation precludes the use of the filtration method for the quantification of protozoan grazing in a community where a high concentration of particles is present, and where a large portion of the active microbial community would be expected to be associated with those particles.

3. Dilution

The dilution technique assumes that observed bacterial growth in a diluted water sample is quantitatively equivalent to the grazing rate in unaltered samples.[7,61,62] Dilution of the water sample with filtered water lowers the concentration of grazers and bacteria (removed from the diluent during filtration). Protozoan grazing rates are calculated from the difference between the apparent bacterial growth rates in the diluted and the control treatments, which are then corrected by the respective dilution factor. The underlying assumptions for the

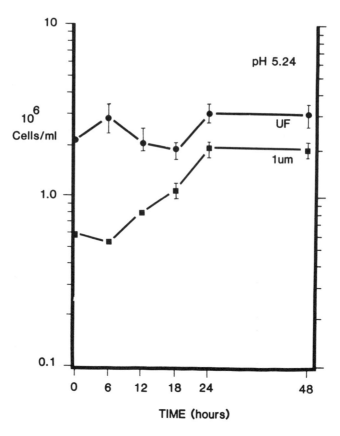

FIGURE 5.   Estimation of protozoan grazing on bacteria at pH 5.24 as determined by the filtration method. Actual grazing rates were calculated on the basis of the regression of lnY vs. X for the 1st 24 h of the incubation. UF (●) represents unfiltered controls; 1 μm (■) represents suspensions that had been passed through a 1-μm pore size filter. Error bars represent 1 SE of the mean.

dilution method are (1) that bacterial growth follows an exponential model, (2) that the change in bacterial growth rate is directly proportional to the dilution factor, and (3) that bacterial growth is unaltered by inadvertent substrate enrichment. These assumptions were examined in a series of experiments which have been reported on elsewhere,[63] and a summary of those findings is given here. First, the appropriate model (linear vs. exponential) was selected based on statistical analyses of the residuals from a least squares regression of bacterial concentration over time. We found for our system, the Lake Anna sediment-water interface, that the exponential model described bacterial growth in diluted waters. Second, an ANOVA with a lack-of-fit test, determined that the change in bacterial growth rate was directly proportional to the dilution factor. Finally, using ANOVA in two carbon enrichment experiments, we demonstrated that dilution had a significant effect on the estimate of apparent bacterial growth rate while carbon addition did not. These results indicate that the assumptions are not violated and the dilution method provides a good estimate of grazing mortality. Grazing rates determined by this method at Lake Anna range between 0.018 and 0.115 h^{-1}.

4. Summary

A review of the methods used to quantify protozoan grazing on bacteria in aquatic ecosystems has shown that (1) there are severe limitations with the use of eukaryote and prokaryote inhibitors and (2) that the use of labeled particles, particularly fluorescent mi-

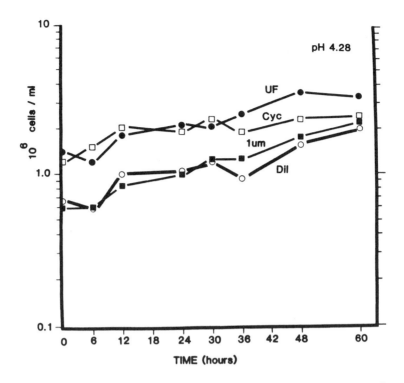

FIGURE 6. Comparison of protozoan grazing on bacteria with time as measured using several different methods. UF (●) represents unfiltered, control suspensions; Cyc (□) indicates inhibition of eukaryotic cells with cycloheximide (100 mg l^{-1}); 1μm (■) represents suspensions passed through a 1-μm pore size filter; and Dil (○) represents suspensions diluted with filtered lake water (3:1 dilution). Each data point is the log of the mean bacterial concentration for three replicate treatment flasks. The SE of the means were usually <10% and are not plotted for clairity. Grazing rates were calculated from the regression of lnY vs. X for the 1st 24 h of the incubation.

Table 8
COMPARISON OF LITERATURE VALUES WITH ESTIMATES OF PROTOZOAN GRAZING RATE AT THE SEDIMENT-WATER INTERFACE OF LAKE ANNA, VA, USING INHIBITION, FILTRATION, AND DILUTION

Method	Protozoan grazing rate (h^{-1})	Ref.
Inhibition	0.01—0.03	43, 45, 46
	0	This study
Filtration	0.02—0.12	60
	0.06—0.11	This study
Dilution	0.02—0.07	61, 62
	0.02—0.11	This study

crospheres, may be limited by the effect of grazing bias. Our review has also shown that the filtration and dilution methods remain viable alternatives when the limitations and assumptions are carefully evaluated.

We conducted two experiments where we used three of the methods: eukaryote inhibition (cycloheximide), 1-μm filtration, and 1:3 dilution (Figure 6). The results of a similar experiment has been reported elsewhere,[52] but briefly, bacterial growth in the cycloheximide treatment was not significantly different than that in the untreated control and demonstrated the failure of eukaryote inhibition in a field sample. Bacterial growth rates in filtration and dilution treatments were not significantly different from each other, but were significantly different from bacterial growth rates in the control. Protozoan grazing rates calculated from these two treatments were quite similar, 0.019 and 0.016 h^{-1} for filtration and dilution methods, respectively. Examining all published values for grazing mortality and bacterial production using the methods discussed here reveals that values provided by these methods are similar, spanning only an order of magnitude (Table 8). This observation is quite remarkable given the limitations of the methods and variations in temperature, latitude, productivity, etc.

Given all of those constraints and limitations, we will now consider what is known of the effect of acidification and acid stress on protozoan bacterivory.

III. RESPONSE OF PROTOZOANS TO ACIDIFICATION

A. Water Column
1. Food Chain Studies

The response of food chains to acidification has been well documented. Acidification generally results in loss of fish populations, causing food chain readjustments whereby invertebrates become the dominant predators, with associated changes in the grazer, bacterial, and algal communities.[64-68] Fish disappear from acidified lakes as a result of (1) direct toxicity of acid, (2) disruption of reproduction and recruitment, (3) disruption of the food chain, (4) disappearance of spawning sites, and (5) toxic effects of increased soluble aluminum.[69-71] Acidification of lakes and rivers in North America has resulted in the disappearance of fish in at least 200 Adirondack lakes and 200 Ontario lakes; the mechanism is most often the failure of recruitment of new age classes into the population due to postembryonic mortality.[70-71] Benthic invertebrates, especially mollusks, are also susceptible to acidification. Elimination of the top predators (fish) is followed by an increased abundance of invertebrate predatory aquatic insects (e.g., *Chaoborus* spp., littoral odonate larvae) and an observed shift in zooplankton community structure (e.g., a shift from cladoceran dominance to copepod dominance or the domination of the zooplankton by a few species of large cladocerans). This shift may be due to one or all of the following: (1) direct acid sensitivity of some cladoceran species,[72] (2) increasing abundance of invertebrate predator species which preferentially consume cladocerans,[73] and/or (3) removal of fish grazing pressure on zooplankton.[64,65,74-76] In a study of a nonacidified Danish lake, the presence of fish resulted in zooplankton (50 to 140 μm) consuming 4% of bacterial net secondary production and the majority of bacterial production is available to the 5×10^3 flagellates per milliliter. When fish were removed, zooplankton consumed 48 to 51% of net bacterial production and flagellate abundance was reduced to 0.5×10^2 to 2.3×10^2 per milliliter.[77]

At the base of the food chain, both phytoplankton abundance (chlorphyll-a concentration) and the number of taxa are reduced in acidified lakes as compared to natural lakes.[65] The structure of the phytoplankton community changes shifting from predominantly Crysophyte and blue-green species to Chlorophyte (green algae) species with increasing acidification.[65,69]

The picture is not as clear for the nonphotosynthetic bacterioplankton where there is considerable disagreement. Some studies report no difference in bacterial abundance when comparing acidified vs. nonacidified systems or stations within systems using the direct count method;[79-81] other researchers report a significant difference in bacterial abundance. Rao and Dutka[82] found concentrations of bacteria were an order of magnitude less in acidic vs. nonacidic lakes in the Sudbury area, but the additional stress of high metal loadings in the acidified lakes studies may account for this disagreement with other reports.

Reports on the effect of acidification on bacterial productivity are equivocal. Decomposition rates (using litter bags) and heterotrophic activity have been reported to be depressed[81] or not significantly different[80] under acid stress relative to those rates reported for nonacidified sites. The decomposition rates for cellulose strips were not significantly different in the water column of an acid vs. a nonacid lake.[79] Palumbo et al.[83] measured bacterial productivity in acidic vs. nonacidic streams in the Appalachian Mountains with the ^{3}H-thymidine method and found that the acid status of the streams made little difference. The presence of aluminum, however, seemed to inhibit production of bacteria in the leaf material.

In summary, although bacterial concentration is not lower in the water column of acidified systems, bacterial decomposition rates may be lower in highly acidified situations such as acid mine drainage-polluted environments, but the activity of bacteria in sites acidified by precipitation is less effected.

2. Lake Survey Studies
a. Introduction

The response of protozoans to pH has been considered important since the 1929 study of seasonal protozoan distribution by Wang.[84] He observed that several species of protozoans disappeared at a pH of 6.2 when compared to more neutral pH levels (pH 7.05). Since then, Ehrlich[85] found a flagellate and an amoeba in acid mine drainage from a copper mine at a pH of 2.5. The flagellate was carried through three transfers in an inorganic salts medium specific for *Thiobacillus ferrooxidans* (9K medium) along with the bacteria. These observations serve to show that protozoans can live, feed, and reproduce in low pH conditions.

Research has progressed from these observations to complex multivariate analyses of protozoan distribution and the relation to physiochemical and trophic parameters. There have been only three major studies which provide information on protozoan-pH relationships in the water column of freshwater ecosystems: the work of Beaver and Crisman,[22,23] who studied the biotic response to acidification in over 20 Florida lakes, Gates and Lewg,[24] who studied 9 oligotrophic lakes in Ontario, and the research of Cairns and co-workers[20,25,26] who studied protozoans in 5 wetland types in northern lower Michigan.

b. Florida Lakes

Beaver and Crisman[22] reported that ciliate abundance and biomass decreased with decreasing pH (Table 9). Additionally, the community composition shifted to species of smaller size (20 to 30 fm length) with decreasing pH. Food availability appeared to be the primary control on ciliate biomass and abundance in the 20 lakes studied (i.e., decreased bacterial productivity at decreased pH as discussed above). Beaver and Crisman[23] also evaluated the relationship between ciliate biomass and lake trophic status (using Carlson's[86] Trophic State Index). Ciliate biomass and abundance increased with increasing trophic status (Table 10). Note that mean pH was similar across lake trophy. Again, increased food availability with increasing lake trophy was the underlying mechanism for the relationship between lake trophy and ciliate biomass and abundance.

c. Ontario Lakes

Gates and Lewg[24] reported that in nine oligotrophic Ontario lakes, ciliate standing crop (concentration) was significantly correlated with total organic carbon and total inorganic

Table 9
CILIATE BIOMASS AND ABUNDANCE OF 21 FLORIDA
LAKES IN pH CLASSES, WITH THE NUMBER (N) OF
LAKES IN EACH pH CLASS

pH class	N	Ciliate biomass (milligrams C per cubic meter)	Ciliate abundance (number per milliliter)
<5.0	4	3.8 ± 10.4	3.7 ± 0.6
5.0—5.5	7	13.3 ± 8.9	9.3 ± 4.8
5.5—6.0	2	20.8 ± 20.3	17.6 ± 14.6
6.0—7.0	8	16.8 ± 10.4	23.1 ± 11.7

Data from Beaver, J. R. and Crisman, T. L., *Verh. Int. Ver. Limnol.*, 21, 353, 1981. With permission.

Table 10
CILIATE BIOMASS AND ABUNDANCE OF 20 FLORIDA LAKES
SORTED BY LAKE TROPHY

Trophy	N	Mean pH	Ciliate biomass (milligrams C per cubic meter)	Ciliate abundance (number per milliliter)
Oligotrophic	6	5.96	9.3 ± 5.6	10.8 ± 5.4
Mesotrophic	8	5.72	25.1 ± 11.4	29.5 ± 7.7
Eutrophic	2	4.98	26.0 ± 2.9	55.5 ± 7.6
Hypereutrophic	4	5.36	126.0 ± 89.2	155.0 ± 60.9

Note: The number (N) of lakes and the mean pH in each trophic class are given.

Data from Beaver, J. R. and Crisman, T. L., *Limnol. Oceanogr.*, 27, 246, 1982. With permission.

carbon. Ciliate biomass (grams of carbon) was significantly correlated with total phosphorus, conductivity, Kjeldahl nitrogen, and total inorganic carbon. In this study, the authors suggested that variations in ciliate biomass and abundance were most directly linked to bacterial productivity rather than the physical-chemical parameters described above (including pH), as the ciliates were the primary bacterivores in the plankton and the bacteria (food supply) was controlled by the abiotic parameters measured.

d. Michigan Wetlands

The five wetland types studied by the Cairns group were bog, swamp, fen, marsh, and lake; the mean pH values for the wetland types were, 4.7, 6.7, 6.9, 7.5, and 8.2, respectively. The research on protozoans in these wetland lakes is different from the Florida and Ontario studies in that natural protozoan communities were not enumerated; rather, the authors enumerated protozoans which colonized artificial substrates made of polyurethane foam (PF blocks). The data were reported as colonization rates.

Protozoan colonization rates correlated with primary productivity (r = 0.850, P = 0.001) and water color (r = 0.599, P = 0.084).[25] The bog lakes were highly colored, acidic, had the highest primary productivity, and had the highest chlorophyll-a, phosphorus, and nitrogen concentrations. From these parameters it was concluded that bogs were the most eutrophic systems studied, which contradicts previous designations of dystrophy. Protozoan colonization rates were high in the acidic bog lakes as compared to the other systems. The colonization rate by day 1 (i.e., the mean percentage of the maximum number of species over the whole time period, which colonized the PF blocks by day 1) for each lake type are as follows: Douglas Lake, 29.9%; swamp, 39.4%; marsh, 46.9%; bog, 55.4%; and fen, 81.5%, which indicates that pH did not control this index of colonization rate.

The data suggest a change in species composition with wetland type and perhaps pH. Most sites had approximately equal numbers of ciliates and flagellates with the exception of the bogs, where three to five times as many flagellates as ciliates were found. The percentage of the total species collected which were bacterivores ranged from 60.5 to 73.4% across the woodland types, while the percentages that were photosynthetic ranged from 16.0 to 31.4%. Although there is no direct evidence of an alteration in community structure as a response to acidification, such a response might be expected based on experience with other microbes.

In another study of six wetlands (a subset of those studied above plus a river), high particulate organic carbon (POC) concentrations correlated with greatest number of species of protozoans having colonized the PF blocks.[20] There were changes in the composition of the colonizing community over the year of the study, but year-to-year stability in the number of species collected. All of the protozoan communities recovered were dominated by bacterivores and detritivores, which suggested that the detrital food webs are especially important in these environments, where high concentrations of particulate organic carbon allows for high bacterial production and are, in turn, available to protozoa.

Finally, Stewart et al.[26] reported on a multivariate statistical analysis of three lakes, a bog, and a marsh. In this study, pH, conductivity, alkalinity, chloride, silica, temperature, dissolved oxygen, ammonia, total phosphorus, and ortho-phosphate were measured in addition to protozoan colonization of the PF blocks. The analyses indicated that protozoan samples from the five ecosystems appeared to be separated primarily by pH and dissolved oxygen concentrations (this latter factor will be discussed in greater detail below).

e. Summary

All three of these major studies of acidified systems have concluded that POC concentrations and bacterial productivity are the controlling factors affecting protozoan distribution. The acidity of the water does not strictly limit the abundance of protozoa. Cairns' group showed that the total number of species which colonized the PF blocks and the percentage of the maximum number of species which had colonized by day 1 were not pH dependent (i.e., species richness and rate of colonization were not pH dependent) (Figure 7). We have plotted ciliate biomass vs. pH for the Florida lakes which did not have colored water (color <100 Pt units)[22,87] (Figure 8). A least squares regression of ciliate biomass vs. pH did not give a significant regression ($Y = -40.9 + 15.34 X$; $N = 19$; $r = 0.34$). This observation is in disagreement with that of the authors publishing the data as discussed earlier (Section b), who used pH classes and concluded a relationship was present (Table 9). Gates and Lewg[24] reported pH values and ciliate biomass values (Figure 9). Again, a least squares regression of ciliate biomass vs. pH did not give a significant regression ($Y = -50.7 + 17.20 X$; $N = 9$; $r = 0.51$).

B. Sediments

1. The Chemical Environment

The pH of sediment pore waters is significantly altered by bacterial processes.[88] pH values can be changed from acidic levels in the water column of 3.5 to pore water pH values of 6.5 to 7.0 within 1 or 2 cm of the sediment surface.[89-91] Sediment oxygen concentrations vary greatly from lake to lake, and usually depend on lake stratification. Increased sulfate loading from acidified inputs (acid mine drainage or acid rain) can stimulate bacterial sulfate reduction; concentrations of hydrogen sulfide in anoxic water or pore water may be higher than that in lakes not receiving additional sulfate loading. Lakes receiving heavy metals in

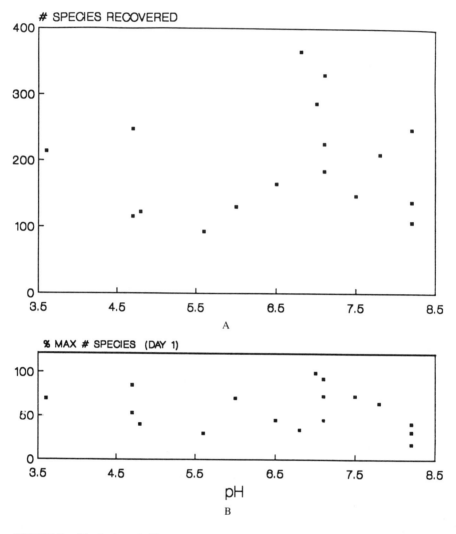

FIGURE 7.   Distribution of ciliate colonization rate variables in wetland of differing pH. (A) Distribution of total number of species recovered from a wetland using poly foam (PF) blocks vs. pH. (B) The mean percentage of the total number of species which colonized the PF blocks by the 1st d vs. pH. (Data from Henebry, M. S. and Cairns, J., Jr., *J. Protozool.*, 31, 456, 1984. With permission.)

association with the acid pollution have higher concentrations of toxic, reduced, and/or volatile metal compounds in the sediment pore waters. Metal loading may create a distinctly different class of lakes where the impact of acidification of the biota is significantly increased by the impact of toxic metals.[76]

2. Bacteria in Sediments of Acidified Lakes

Total bacterial abundance is not significantly different in the sediments of acidified lakes;[78,79] however, some functional groups of bacteria (e.g., nitrifiers) are found in reduced numbers in acid systems.[79,82] *In situ* decomposition in sediments is unaffected by epilimnetic pHs ranging down to 5.1, as measured by methane production, dissolved inorganic carbon production, oxygen uptake, thymidine incorporation, and glucose metabolism.[90,92,93] However, decomposition of cellulose was significantly lower in the sediments of an acidified lake as compared to a nonacid lake.[79] As pointed out elsewhere in this volume,[88] sulfate

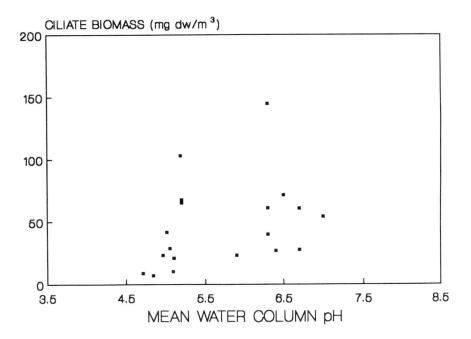

FIGURE 8.   Distribution of ciliate biomass (measured as the milligrams dry weight of ciliates per cubic meter of water) in lakes of differing pH. (Data from Beaver, J. R. and Crisman, T. L., *Verh. Int. Ver. Limnol.*, 21, 353, 1981. With permission.)

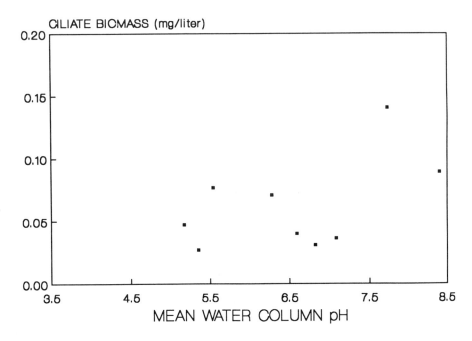

FIGURE 9.   Distribution of ciliate biomass (measured as milligrams wet weight per liter of water) in lakes of differing pH. (Data from Gates, M. A. and Lewg, U. T., *J. Plankton Res.*, 6, 443, 1984. With permission.)

reduction, fermentation, and acetate mineralization rates in acid lake sediments are often very high. Bacterial growth can be strongly inhibited by acidified water containing metals and metalloids (Hg, Pb, As, and Se).[94,95] High concentrations of metals may contribute to the pH-dependent response of respiring bacteria and total organics in the sediments of lakes near Sudbury, Ontario.[82,96]

3. Protozoans in the Benthos

There are no published reports on protozoans in the sediments of acidified lakes. Ciliates are abundant in lake sediments; greatest densities are found at the oxic/anoxic interface. Lakes which stratify have decreasing oxygen concentrations in hypolimnetic waters. Ciliates respond to the upward migration of the oxic/anoxic interface by migrating out of the sediments into the water column.[97-99] Ciliates appear to have maximum densities at oxygen concentration of about 0.5 mg l^{-1} [98] or, in some cases, just below the thermocline.[100,101] The negative response to increasing oxygen concentration shown by *Loxodes* appears to be due to the generation of superoxide in a photosensitive pigment when the organism is exposed to light (in excess of 10 W m^{-2}) while in oxic water.[102,103] The greatest densities of ciliates have been found in sediments from the deepest sites.[104,105] Jones[105] found two to three times the density of ciliates in profundal sediments than in littoral sediments, and Goulder[104] offered three reasons in explanation: increased food availability, decreased turbulence, and increased interstitial space in profundal sediments as compared with the nearshore locations. Using time series analysis, hypolimnetic ciliate biomass was determined to be correlated to the bacterial biomass 2 weeks prior to ciliate quantification ($r^2 = 0.89$) and organic carbon sedimentation 4 weeks prior to ciliate quantification ($r^2 = 0.85$).[106] The consistent lags observed in this study underscore the close relationship between protozoan biomass and food availability as discussed earlier.

C. The Sediment-Water Interface

1. The Chemical and Physical Environment

The sediment-water interface in lakes has the potential for high biological activity in lakes which are at least partially oxygenated throughout the bottom waters. In the context of acidified systems with an oxygenated water column, protozoan activity which might be minimal in the water column due to limited food resources would be enhanced at the sediment-water interface. Rao and Dutka[82] reported that organic carbon concentrations in the sediments of acidified systems are three to four times that found in nonacid systems. This could be due to (1) increased carbon loading to the sediments or (2) decreased bacterial mineralization of the organic material; this question remains unanswered. Some researchers report high bacterial activity in the sediments of acidified lakes (see Reference 88). Protozoans living at the sediment-water interface have abundant food supplies and predator refugia available from the sediments while being able to adjust their vertical position in the interface in accordance with their redox requirements.

2. Protozoan Grazing at the Sediment-Water Interface

We have conducted 12 protozoan grazing experiments at the sediment-water interface at acidified and nonacidified stations of Lake Anna, VA, using the dilution method. Acid mine drainage enters one arm of Lake Anna where physical and biological processes act to increase the pH and reduce metal, sulfate, and iron concentrations.[88,107,108] An adjacent arm of the lake (Freshwater Creek), which does not receive acid mine drainage, is the study site for nonacidified samples. Marginal freshwater marshes and submerged aquatic vegetation provide substantial carbon sources to support microbial activities even in the acidified arm of the lake.[109] Protozoan grazing rates were calculated as described previously,[63] and have been plotted vs. pH (Figure 10). Protozoan grazing rates were not significantly affected by pH,

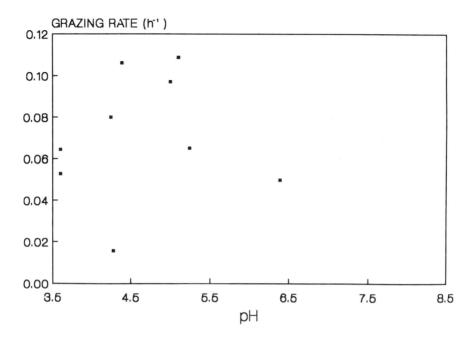

FIGURE 10.   Distribution of protozoan grazing rate (grams) with pH in Lake Anna.

nor was the total number of bacteria grazed. The grazing rates obtained in the acidic areas of Lake Anna compare favorably with those reported for other aquatic systems (Table 8).

IV. CONCLUSIONS

Clearly, data on protozoan grazing rates in acidified lakes are inadequate. We believe, however, that from the ecosystem point of view, the data collected thus far, along with the research discussed in earlier sections of this chapter, indicate that pH does not directly affect protozoans or their activities, that is, the species present may be different at various pH values, but the function of the community would appear to remain intact. The indirect effects of acidification on algal productivity and, in turn, the supply of organic matter to bacteria is an important controlling factor in bacterivorous protozoan distribution. Adequate organic substrate for high rates of bacterial productivity may be available in sediments, allowing for abundant protozoan communities to develop in the benthos when they cannot develop in the water column. Research into the synergistic effects of metals and acid on sediment protozoan communities could clarify the effect of these confounded variables.

REFERENCES

1. **Ramirez, C. and Alexander, M.,** Evidence suggesting protozoan predation on *Rhizobium* associated with germinating seeds in the rhizosphere of beans (*Phaseolus vulgaris* L.), *Appl. Environ. Microbiol.,* 40, 492, 1980.
2. **Ingham, E. R. and Coleman, D. C.,** Effects of streptomycin, cycloheximide, Fungizone, captan, car-bofuran, cygon, and PCNB on soil microorganisms, *Microb. Ecol.,* 10, 345, 1984.
3. **Hendrix, P. F., Parmelee, R. W., Crossley, D. A., Jr., Coleman, D. C., Odum, E. P., and Groffman, P. M.,** Detritus food webs in conventional and no-tillage agroecosystems, *Bioscience,* 36, 374, 1986.
4. **Lighthart, B.,** Planktonic and benthic bacterivorous protozoa at eleven stations in Puget Sound and adjacent Pacific Ocean, *J. Fish. Res. Board Can.,* 26, 299, 1969.

5. **Fenchel, T. M. and Jorgensen, B. B.,** Detritus food chains of aquatic ecosystems: the role of bacteria, in *Advances in Microbial Ecology,* Vol. 1, Alexander, M., Ed., Plenum Press, New York, 1977, 1.

6. **Sherr, B. F. and Sherr, E. B.,** Enumeration of heterotrophic microprotozoa by epifluoresence microscopy, *Estuarine Coastal Shelf Sci.,* 16, 1, 1983.

7. **Wright, R. T. and Coffin, R. B.,** Ecological significance of biomass and activity measurements, in *Current Perspectives in Microbial Ecology,* Klug, M. J. and Reddy, C. A., Eds., American Society for Microbiology, Washington, D.C., 1984, 485.

8. **Pomeroy, L. R.,** The ocean's food web: a changing paradigm, *Bioscience,* 24, 499, 1974.

9. **Porter, K. G., Sherr, E. B., Sherr, B. F., Pace, M. L., and Sanders, R. W.,** Protozoa in planktonic food webs, *J. Protozool.,* 32, 409, 1985.

10. **Fuhrman, J. A. and Azam, F.,** Bacterioplankton secondary production estimates for coastal waters of British Columbia, Antarctica, and California, *Appl. Environ. Microbiol.,* 39, 1085, 1980.

11. **Rheinheimer, G.,** Investigations on the role of bacteria in the food web of the Western Baltic, *Kiel. Meeresforsch. Sonderh.,* 5, 284, 1981.

12. **Chrost, R. J.,** The composition and bacterial utilization of DOC released by phytoplankton, *Kiel. Meeresforsch. Sonderh.,* 5, 325, 1981.

13. **Wetzel, R. G.,** *Limnology,* 2nd ed., Saunders College, New York, 1983.

14. **Hargrave, B. T.,** The effect of a deposit-feeding amphipod on the metabolism of benthic microfauna, *Limnol. Oceanogr.,* 15, 21, 1970.

15. **Starkweather, P. L., Gilbert, J. J., and Frost, T.,** Bacterial feeding by the rotifer *Brachyonos calyciflorus:* clearance and ingestion rates, behavior and population dynamics, *Oecologia,* 44, 26, 1979.

16. **Newell, R.,** The role of detritus in the nutrition of two marine deposit feeders, the prosobranch *Hydrobia ulvae* and the bivalve *Macoma balthica, Proc. Zool. Soc. London.,* 144, 24, 1965.

17. **Kofoed, L. H.,** The feeding biology of *Hydrobia ventrosa* (Monatagu). II. Allocation of the components of the carbon-budget and the significance of the secretion of dissolved organic material, *J. Exp. Mar. Biol. Ecol.,* 19, 243, 1975.

18. **Hicks, G. R. and Coull, B. C.,** The ecology of marine meiobenthic harpacticoid copepods, *Oceanogr. Mar. Biol. Annu. Rev.,* 21, 67, 1983.

19. **ZoBell, C. E. and Landon, W. A.,** The bacterial nutrition of the California mussel, *Proc. Soc. Exp. Biol. Med.,* 36, 607, 1937.

20. **Pratt, J. R. and Cairns, J., Jr.,** Functional groups in the protozoa: roles in differing ecosystems, *J. Protozool.,* 32, 415, 1985.

21. **Azam, F., Fenchel, T., Field, J. G., Gray, J. S., Meyer-Reil, L. A., and Thingstad, F.,** The ecological role of water-column microbes in the sea, *Mar. Ecol. Prog. Ser.,* 10, 257, 1983.

22. **Beaver, J. R. and Crisman, T. L.,** Acid precipitation and the response of ciliated protozoans in Florida lakes, *Verh. Int. Ver. Limnol.,* 21, 353, 1981.

23. **Beaver, J. R. and Crisman, T. L.,** The trophic response of ciliated protozoans on freshwater lakes, *Limnol. Oceanogr.,* 27, 246, 1982.

24. **Gates, M. A. and Lewg, U. T.,** Contribution of ciliated protozoa to the planktonic biomass in a series of Ontario lakes: quantitative estimates and dynamical relationships, *J. Plankton Res.,* 6, 443, 1984.

25. **Henebry, M. S. and Cairns, J., Jr.,** Protozoan colonization rates and trophic status of some freshwater wetland lakes, *J. Protozool.,* 31, 456, 1984.

26. **Stewart, P. M., Smith, E. P., Pratt, J. R., McCormick, P. V., and Cairns, J., Jr.,** Multivariate analysis of protist communities in lentic systems, *J. Protozool.,* 33, 152, 1986.

27. **Sieracki, M. E., Hass, L. W., Caron, D. A., and Lessard, E. J.,** submitted.

28. **Sanders, R. W.,** personal communication.

29. **Jonsson, P. R.,** Particle size selection, feeding rates and growth dynamics of marine planktonic oligotrichous ciliates (Ciliophora: Oligotrichina), *Mar. Ecol. Prog. Ser.,* 33, 265, 1986.

30. **Bird, D. F. and Kalff, J.,** Bacterial grazing by planktonic lake algae, *Science,* 231, 493, 1986.

31. **McManus, G. B. and Fuhrman, J. A.,** Bacterivory in seawater studied with the use of inert fluorescent particles, *Limnol. Oceanogr.,* 31, 420, 1986.

32. **Sherr, B. F., Sherr, E. B., and Fallon, R. D.,** Use of monodispersed, fluorescently labeled bacteria to estimate in situ protozoan bacterivory, *Appl. Environ. Microbiol.,* 53, 958, 1987.

33. **Tremaine, S. C. and Mills, A. L.,** unpublished data.

34. **Finlay, B. J.,** Procedures for the isolation, cultivation, and identification of protozoa, in *Experimental Microbial Ecology,* Burns, R. G. and Slater, J. H., Eds., Blackwell Scientific, St. Louis, 1982, 44.

35. **Caron, D. A.,** Technique for enumeration of heterotrophic and phototrophic nanoplankton, using epifluorescence microscopy, and comparison with other procedures, *Appl. Environ. Microbiol.,* 46, 491, 1983.

36. **Hobbie, J. E., Daley, R. J., and Jasper, S.,** Use of Nuclepore filters for counting bacteria by fluorescence microscopy, *Appl. Environ. Microbiol.,* 33, 1225, 1977.

37. **Taylor, G. T., Karl, D. M., and Pace, M. L.,** Impact of bacteria and zooflagellates on the composition of sinking particles: and in situ experiment, *Mar. Ecol. Prog. Ser.,* 29, 141, 1986.

38. **Lessard, E. J. and Swift, E.,** Species-specific grazing rates of heterotrophic dinoflagellates in oceanic waters, measured with a dual-label radioisotope technique, *Mar. Biol.,* 87, 289, 1985.
39. **Li, K. W. and Dickie, P. M.,** Metabolic inhibition of size-fractionated marine plankton radiolabeled with amino acids, glucose, bicarbonate, and phosphate in light and dark, *Microb. Ecol.,* 11, 11, 1985.
40. **Ducklow, H. W., Purdie, D. A., Williams, P. J. LeB., and Davies, J. M.,** Bacterioplankton: a sink for carbon in a coastal marine plankton community, *Science,* 232, 865, 1986.
41. **Ducklow, H. W., Purdie, D. A., Williams, P. J. LeB., and Davies, J. M.,** Bacteria: link or sink?, *Science,* 235, 88, 1987.
42. **Sanders, R. W., Porter, K. G., Bennett, S. J., and DeBaise, A. E.,** submitted.
43. **Fuhrman, J. A. and McManus, G. B.,** Do bacteria-sized marine eukaryotes consume significant bacterial production?, *Science,* 224, 1257, 1984.
44. **McCambridge, J. and McMeekin, T. A.,** Relative effects of bacterial and protozoan predators on survival of *Escherichia coli* in estuarine water samples, *Appl. Environ. Microbiol.,* 40, 907, 1980.
45. **Sanders, R. W. and Porter, K. G.,** Use of metabolic inhibitors to estimate protozooplankton grazing and bacterial production in a monomictic eutrophic lake with an anaerobic hypolimnion, *Appl. Environ. Microbiol.,* 52, 101, 1986.
46. **Sherr, B. F., Sherr, E. B., Andrew, T. L., Fallon, R. D., and Newell, S. Y.,** Trophic interactions between heterotrophic protozoa and bacterioplankton in estuarine water analyzed with selective metabolic inhibitors, *Mar. Ecol. Prog. Ser.,* 32, 169, 1986.
47. **Taylor, G. T. and Pace, M. L.,** Validity of eucaryote inhibitors for assessing production and grazing mortality of marine bacterioplankton, *Appl. Environ. Microbiol.,* 53, 9, 1987.
48. **Krempin, D. W., McGrath, S. M., SooHoo, J. B., and Sullivan, C. W.,** Orthophosphate uptake by phytoplankton and bacterio-plankton from the Los Angeles harbor and southern California coastal waters, *Mar. Biol.,* 64, 23, 1981.
49. **Habte, M. and Alexander, M.,** Protozoan density and the coexistence of protozoan predators and bacterial prey, *Ecology,* 59, 140, 1978.
50. **Campbell, L. and Carpenter, E. J.,** Estimating the grazing pressure of heterotrophic nanoplankton on *Synechococcus* spp. using the seawater dilution and selective inhibitor techniques, *Mar. Ecol. Prog. Ser.,* 33, 121, 1986.
51. **Ennis, H. L. and Lubin, M.,** Cycloheximide: aspects of inhibition of protein synthesis in mammalian cells, *Science,* 146, 1474, 1964.
52. **Tremaine, S. C. and Mills, A. L.,** Inadequacy of the eucaryote inhibitor cycloheximide in studies of protozoan grazing on bacteria at the freshwater sediment-water interface, *Appl. Environ. Microbiol.,* 53, 1969, 1987.
53. **Iturriaga, R. and Zsolnay, A.,** Differentiation between auto- and heterotrophic activity: problems in the use of size fractionation and antibiotics, *Bot. Mar.,* 24, 399, 1981.
54. **Newell, S. Y., Sherr, B. F., Sherr, E. B., and Fallon, R. D.,** Bacterial response to presence of eukaryote inhibitors in water from a coastal marine environment, *Mar. Environ. Res.,* 10, 147, 1983.
55. **Yetka, J. E. and Wiebe, W. J.,** Ecological application of antibiotics as respiratory inhibitors of bacterial production, *Appl. Microbiol.,* 28, 1033, 1974.
56. **Smith, K. L., Jr., Burns, K. A., and Teal, J. M.,** In situ respiration of benthic communities in Castle Harbor, Bermuda, *Mar. Biol.,* 12, 196, 1972.
57. **Joklik, W. K., Willell, H. P., and Amos, D. B.,** *Zinsser Microbiology,* 17th ed., Appleton-Century-Crofts, New York, 1980.
58. **Briody, B. A.,** *Microbiology and Infectious Disease,* McGraw-Hill, New York, 1984, chap. 4.
59. **Cynar, F. J., Estep, K. W., and Sieburth, J. McN.,** The detection and characterization of bacteria-sized protists and "protist-free" filtrates and their potential impact on experimental marine ecology, *Microb. Ecol.,* 11, 281, 1985.
60. **Wright, R. T. and Coffin, R. B.,** Measuring microzooplankton grazing on planktonic marine bacteria by its impact on bacterial production, *Microb. Ecol.,* 10, 137, 1984.
61. **Landry, M. R. and Hassett, R. P.,** Estimating the grazing impact of marine micro-zooplankton, *Mar. Biol.,* 67, 283, 1982.
62. **Ducklow, H. W. and Hill, S. M.,** The growth of heterotrophic bacteria in the surface waters of warm core rings, *Limnol. Oceanogr.,* 30, 239, 1985.
63. **Tremaine, S. C. and Mills, A. L.,** Tests of the critical assumptions of the dilution method for estimating bacterivory by microeucaryotes, *Appl. Environ. Microbiol.,* 53, 2914, 1987.
64. **Stenson, J. A. E. and Oscarson, H. G.,** Crustacean zooplankton in the acidified Lake Gardson system, *Ecol. Bull.,* 37, 224, 1985.
65. **Crisman, T. L., Schulze, R. L., Brezonik, P. L., and Bloom, S. A.,** Acid precipitation: the biotic response in Florida lakes, in *Proc. Int. Conf. Ecological Impact of Acid Precipitation,* Drablos, D. and Tollan, A., Eds., SNSF Project, Oslo, 1980, 296.

66. **Hessen, D. O. and Nilssen, J. P.,** From phytoplankton to detritus and bacteria: effects of short-term nutrient and fish perturbations in a eutrophic lake, *Arch. Hydrobiol.,* 105, 273, 1986.

67. **Nilssen, J. P.,** Acidification of a small watershed in southern Norway and some characteristics of acidic aquatic environments, *Int. Rev. Gesamten Hydrobiol.,* 65, 177, 1980.

68. **Riemann, B.,** Potential importance of fish predation and zooplankton grazing on natural populations of freshwater bacteria, *Appl. Environ. Microbiol.,* 50, 187, 1985.

69. **Schindler, D. W., Mills, K. H., Malley, D. F., Findlay, D. L., Shearer, J. A., Davies, I. J., Turner, M. A., Linsey, G. A., and Cruikshank, D. R.,** Long-term ecosystem stress: the effects of years of experimental acidification on a small lake, *Science,* 228, 1395, 1985.

70. **Harvey, H. H.,** Widespread and diverse changes in the biota of North American lakes and rivers coincident with acidification, in *Ecological Impact of Acid Precipitation,* Drablos, D. and Tollan, A., Eds., SNSF Project, Oslo, 1980, 93.

71. **Muniz, I. P. and Leivestad, H.,** Acidification — effects on freshwater fish, in *Proc. Int. Conf. Ecological Impact of Acid Precipitation,* Drablos, A. and Tollan, A., Eds., SNSF Project, Oslo, 1980, 84.

72. **Nilssen, J. P.,** An ecological jig-saw puzzle: reconstructing aquatic biogeography and pH in an acidified region, *Inst. Freshwater Res., Drottningholm,* 61, 138, 1984.

73. **Nyman, H. G., Oscarson, H. G., and Stenson, J. A. E.,** Impact of invertebrate predators on the zooplankton composition in acid forest lakes, *Ecol. Bull.,* 37, 239, 1985.

74. **Harvey, H. H. and McArdle, J. M.,** Composition of the benthos in relation to pH in the LaCloche lakes, *Water Air Soil Pollut.,* 30, 529, 1986.

75. **Malley, D. F. and Chang, P. S. S.,** Increase in the abundance of cladocera at pH 5.1 in experimentally-acidified lake 223, Experimental Lakes Area, Ontario, *Water Air Soil Pollut.,* 30, 629, 1986.

76. **Yan, N. D. and Strus, R.,** Crustacean zooplankton communities of acidic, metal-contaminated lakes near Sudbury, Ontario, *Can. J. Fish. Aquat. Sci.,* 37, 2282, 1980.

77. **Riemann, B.,** Potential importance of fish predation and zooplankton grazing on natural populations of freshwater bacteria, *Appl. Environ. Microbiol.,* 50, 187, 1985.

78. **Wassel, R. A. and Mills, A. L.,** Changes in water and sediment bacterial community structure in a lake receiving acid mine drainage, *Microb. Ecol.,* 9, 155, 1983.

79. **Hoeniger, J. F. M.,** Decomposition studies in two central Ontario lakes having surficial pHs of 4.6 and 6.6, *Appl. Environ. Microbiol.,* 52, 489, 1986.

80. **Traaen, T. S.,** Effects of acidity on decomposition of organic matter in aquatic environments, in, *Proc. Int. Conf. Ecological Impact of Acid Precipitation,* Drablos, A. and Tollan, A., Eds., SNSF Project, Oslo, 1980, 340.

81. **Carpenter, J., Odum, W. E., and Mills, A. L.,** Leaf litter decomposition in a reservoir affected by acid mine drainage, *Oikos,* 41, 165, 1983.

82. **Rao, S. S. and Dutka, B. J.,** Influence of acid precipitation on bacterial populations in lakes, *Hydrobiologia,* 98, 153, 1983.

83. **Palumbo, A. V., Mulholland, P. J., and Elwood, J. W.,** Bacterial production and aluminum accumulation on decomposing leaf material in streams, *Abstr. Ann. Meet. Am. Soc. Limnol. Oceanogr.,* 1987.

84. **Wang, C. C.,** Ecological studies of the seasonal distribution of protozoa in a fresh-water pond, *J. Morph. Physiol.,* 46, 431, 1928.

85. **Erlich, H. L.,** Microorganisms in acid drainage from a copper mine, *J. Bacteriol.,* 86, 350, 1963.

86. **Carlson, R. E.,** A trophic state index for lakes, *Limnol. Oceanogr.,* 22, 361, 1977.

87. **Beaver, J. R.,** personal communication.

88. **Mills, A. L., Bell, P. E., and Herlihy, A. T.,** Microbes, sediments and acidified water: the importance of biological buffering, in *Microbial Interactions in Acid-Stressed Aquatic Ecosystems,* Rao, S. S., Ed., CRC Press, Boca Raton, FL, 1989, chap. 1.

89. **Herlihy, A. T. and Mills, A. L.,** The pH regime of sediments underlying acidified waters, *Biogeochemistry,* 2, 377, 1986.

90. **Kelly, C. A., Rudd, J. W. M., Furutani, A., and Schindler, D. W.,** Effects of lake acidification on rates of organic matter decomposition in sediments, *Limnol. Oceanogr.,* 29, 687, 1984.

91. **Mitchell, M. J., Landers, D. H., and Brodowski, D. F.,** Sulfur constituents of sediments and their relationship to lake acidification, *Water Air Soil Pollut.,* 16, 351, 1981.

92. **Gahnstrom, G., Andersson, G., and Fleischer, S.,** Decomposition and exchange processes in acidified lake sediment, in *Proc. Int. Conf. Ecological Impact of Acid Precipitation,* Drablos, A. and Tollan, A., Eds., SNSF Project, Oslo, 1980, 306.

93. **Gyure, R. A., Konopka, A., Brooks, A., and Doemel, W.,** Algal and bacterial activities in acidic (pH 3) strip mine lakes, *Appl. Environ. Microbiol.,* in press.

94. **Baker, M. D., Inniss, W. E., Mayfield, C. I., and Wong, P. T. S.,** Effect of acidification, metals, and metalloids on sediment microorganisms, *Water Res.,* 17, 925, 1983.

95. **Mills, A. L. and Colwell, R. R.,** Microbiological effects of metal ions in Chesapeake Bay water and sediment, *Bull. Environ. Contam. Toxicol.,* 18, 99, 1977.

96. **Rao, S. S., Jurkovic, A. A., and Nriagu, J. O.**, Bacterial activity in sediments of lakes receiving acid precipitation, *Environ. Pollut. Ser. A.*, 36, 195, 1984.
97. **Bark, A. W.**, Studies on ciliated protozoa in eutrophic lakes. I. Seasonal distribution in relation to thermal stratification and hypolimnetic anoxia, *Hydrobiologia*, 124, 167, 1985.
98. **Bark, A. W. and Goodfellow, J. G.**, Studies on ciliated protozoa in eutrophic lakes. II. Field and laboratory studies on the effects of oxygen and other chemical gradients on ciliate distribution, *Hydrobiologia*, 124, 177, 1985.
99. **Finlay, B. J.**, Oxygen availability and seasonal migrations of ciliated protozoa in a freshwater lake, *J. Gen. Microbiol.*, 123, 173, 1981.
100. **Pace, M. L.**, Planktonic ciliates: their distribution, abundance, and relationship to microbial resources in a monomictic lake, *Can. J. Fish. Aquat. Sci.*, 39, 1106, 1982.
101. **Pace, M. L. and Orcutt, J. D.**, The relative importance of protozoans, rotifers, and crustaceans in a freshwater zooplankton community, *Limnol. Oceanogr.*, 26, 822, 1981.
102. **Finlay, B. J. and Fenchel, T.**, Photosensitivity in the ciliated protozoan *Loxodes*: pigment granules, absorption and action spectra, blue light perception, and ecological significance, *J. Protozool.*, 33, 534, 1986.
103. **Fenchel, T. and Finlay, B. J.**, Photobehavior of the ciliated protozoan *Loxodes*: taxic, transient, and kinetic responses in the presence and absence of oxygen, *J. Protozool.*, 33, 139, 1986.
104. **Goulder, R.**, The seasonal and spatial distribution of some benthic ciliated protozoa in Esthwaite water, *Freshwater Biol.*, 4, 127, 1974.
105. **Jones, J. G.**, Some differences in the microbiology of profundal and littoral lake sediments, *J. Gen. Microbiol.*, 117, 285, 1980.
106. **Psenner, R. and Schlott-Idl, K.**, Trophic relationships between bacteria and protozoa in the hypolimnion of a meromictic mesotrophic lake, *Hydrobiologia*, 121, 111, 1985.
107. **Mills, A. L. and Herlihy, A. T.**, Microbial ecology and acidic pollution of impoundments, in *Microbial Processes in Reservoirs*, Gunnison, D., Ed., Dr. W. Junk, Dordrecht, Netherlands, 1985, 169.
108. **Herlihy, A. T., Mills, A. L., Hornberger, G. M., and Bruckner, A. E.**, The importance of sediment sulfate reduction to the sulfate budget of an impoundment receiving acid mine drainage, *Water Resour. Res.*, 23, 287, 1987.
109. **Tremaine, S. C., Bell, P. E., and Mills, A. L.**, unpublished data.
110. **Borsheim, K. Y.**, Clearance rates of bacteria-sized particles by freshwater ciliates, measured with monodispersed latex beads, *Oecologia*, 63, 286, 1984.
111. **Fenchel, T.**, Ecology of heterotrophic microflagellates. II. Bioenergetics and growth, *Mar. Ecol. Prog. Ser.*, 8, 225, 1982.
112. **Pace, M. L.**, Experimental evaluation of the fluorescent microsphere technique for measuring protozoan grazing rates, *Abstr. Ann. Meet. Am. Soc. Limnol. Oceanogr.*, 1985.

INDEX